H.B. Spotton

The Commonly Occurring Wild Plants of Canada and More

Especially of the Province of Ontario

H.B. Spotton

The Commonly Occurring Wild Plants of Canada and More Especially of the Province of Ontario

ISBN/EAN: 9783337859350

Printed in Europe, USA, Canada, Australia, Japan

Cover: Foto ©berggeist007 / pixelio.de

More available books at **www.hansebooks.com**

W. J. Gage & Co.'s Educational Series.

THE
COMMONLY OCCURRING
WILD PLANTS OF CANADA,

AND MORE ESPECIALLY OF

THE PROVINCE OF ONTARIO.

A FLORA FOR THE USE OF BEGINNERS,

BY

H. B. SPOTTON, M.A., F.L.S.

PRIN. BARRIE COLL. INST.,

Author of "The Elements of Structural Botany."

REVISED EDITION.

W. J. GAGE & COMPANY,
TORONTO.

CONTENTS.

Preface to the First Edition................................	v
Preface to the Revised Edition.............................	vii
Abbreviations of Names of Authors........................	viii
How to Use the Key and the Flora.......................	ix
Key to the Orders..	xiii

Flora:—
    Flowering or Phanerogamous Plants...................... 1
        Dicotyledons.. 1
            Angiosperms...................................... 1
                Polypetalous Division........................ 1
                Gamopetalous   "   58
                Apetalous   "   116
            Gymnosperms.................................... 139
        Monocotyledons....................................... 143
            Spadiceous Division............................ 143
            Petaloideous   "   147
            Glumaceous   "   165
    Flowerless or Cryptogamous Plants...................... 169
    Index.. 184

PREFACE TO THE FIRST EDITION.

A few words will not be out of place by way of preface to the List of Common Canadian Plants contained in the following pages. It will be observed that the List is confined to wild plants, the exclusion of cultivated Species having been determined on, partly because of the difficulty of knowing where to stop when an enumeration of them has once been entered upon, and partly because it was thought that, on the whole, more important results would be attained by directing attention exclusively to the denizens of our own woods and fields. What is really desired is, to create among our young people an interest in the Botany of Canada, and it seems not unreasonable to hope that this end may be better attained by placing within their reach some such handy volume as the present, dealing only with such plants as grow spontaneously within our limits.

The great majority of the plants described have been personally examined, and their occurrence verified, by the writer, his observations having been directed to what may fairly be regarded as representative districts of the older Provinces, but special acknowledgments are also due to Prof. Macoun, of the Geological Survey, for the free use of his valuable notes, and other friendly assistance.

Whilst diligence has been exercised that no commonly occurring plant should be omitted, yet it can hardly be that such omissions do not occur, and the writer will be most grateful to any observers into whose hands the List may come, if they will kindly draw his attention to any such defects, so that they may be remedied in subsequent editions.

The Classification and Nomenclature adopted are very nearly those of the Fifth Edition of Dr. Gray's Manual of the Botany of the Northern United States, and the writer most gratefully acknowledges the great assistance he has received from the admirable descriptions in that work.

Except in a very general way, no attempt has been made to define the limits of the range of the various Species, as observations tend to show that the range, in many cases, is undergoing constant alteration from various causes. When, however, a Species has appeared to be confined to a particular locality, mention has been made of that fact, but, as a rule, Species known to be of rare occurrence have been excluded.

Characters considered to be of special importance in the determination of the various Species have been emphasized by the use of italics, and where the Species of a Genus, or the Genera of an Order, are numerous, a system of grouping according to some prominent character has been adopted, so as to reduce the labour of determination as much as possible.

To assist the non-classical student, names which might be mispronounced have been divided and accentuated, the division having no reference whatever to the etymology of the words, but being simply based upon their sound when properly pronounced.

It need hardly be added that the writer's ELEMENTS OF STRUCTURAL BOTANY is designed to be the constant companion of the present Flora, in the hands of the young student, for the explanation of such technicalities as he may not have previously mastered.

BARRIE, November, 1883.

PREFACE TO THE REVISED EDITION

The necessity of casting new plates has been taken advantage of to enlarge and otherwise improve the present List of Common Canadian Wild Plants. Descriptions of about one hundred and seventy additional species have been inserted, and one or two which had a place in the old list have, in the light of more thorough investigation, been struck out.

The principal authority for the additions is Macoun's Catalogue of Canadian Plants, issued by the Geological Survey, in which are included the results of the labours of many assiduous workers. The author has, however, also to thank his numerous correspondents in different parts of the country for their very valuable suggestions.

In the present Edition, it will be observed that the authorities for the names are given, and where these names differ from those in Macoun's Catalogue (which should be in the hands of every student), the synonym in the latter is also given.

BARRIE, February, 1889.

PRINCIPAL ABBREVIATIONS OF NAMES OF AUTHORS CITED IN THE FLORA.

Adans.	for	Adanson.	Lam.	for	Lamarck.
Ait.	"	Aiton.	L'Her.	"	L'Heritier.
Bart.	"	Barton.	Lehm.	"	Lehmann.
Beauv.	"	Palisot de Beauvois.	Lindl.	"	Lindley.
Benth.	"	Bentham.	Michx.	"	Michaux.
Bernh.	"	Bernhardi.	Mill.	"	Miller.
Bigel.	"	Bigelow.	Muhl.	"	Muhlenberg.
Boiss.	"	Boissier.	Nees.	"	Nees von Esenbeck
Borkh.	"	Borkhausen.	Nutt.	"	Nuttall.
Cass.	"	Cassini.	Pers.	"	Persoon.
Cav.	"	Cavanilles.	Poir.	"	Poiret.
DC.	"	De Candolle.	R. Br.	"	Robert Brown
A. DC.	"	Alphonse De Candolle.	Raf.	"	Rafinesque.
			Rich.	"	Richard.
Desf.	"	Desfontaines.	Richards.	"	Richardson.
Dill.	"	Dillenius.	Salisb.	"	Salisbury.
Ell.	"	Elliott.	Schreb.	"	Schreber.
Endl.	"	Endlicher.	Scop.	"	Scopoli.
Engelm.	"	Engelmann.	Spreng.	"	Sprengel.
Gært.	"	Gærtner.	Torr.	"	Torrey.
Griseb.	"	Grisebach.	Tourn.	"	Tournefort
Gronov.	"	Gronovius.	Vaill.	"	Vaillant.
Hoffm.	"	Hoffmann.	Vent.	"	Ventenat.
Hook.	"	W. J. Hooker.	Vill.	"	Villars.
H. B. K.	"	Humboldt, Bonpland, and Kunth.	Wahl.	"	Wahlenberg.
			Walt.	"	Walter.
Jacq.	"	Jacquin.	Wangh.	"	Wangenheim
Juss.	"	Jussieu.	Willd.	"	Willdenow.
L.	"	Linnæus.			

HOW TO USE THE KEY AND THE FLORA.

Assuming that the student has carefully read the Introductory part of this work, and is familiar with the ordinary botanical terms, and the chief variations in plant-structure as there set forth, it should, with the aid of the accompanying Key, be a very simple task to refer to its proper Family any Canadian wild plant of common occurrence. To illustrate the method of using this Key, let us suppose that specimens of the following plants have been gathered, and that it is desired to ascertain their botanical names, that is, the name of the Genus and the Species of each:—Red Clover, Strawberry, Blue Flag, and Cat-tail Flag.

All of these produce flowers of some kind, and must therefore be looked for under the head of FLOWERING, OR PHANEROGAMOUS, PLANTS.

With the specimen of Red Clover in hand, and the book open at page xiii, we find that we have first to determine whether our plant is Dicotyledonous or not. The veining of the leaves suggests that it is so; and this impression is confirmed by the fact that the parts of the flower are in fives. Then, is the plant an ANGIOSPERM? As the seed will be found enclosed in an ovary, we answer—Yes. Has the plant both calyx and corolla? Yes. Are the parts of the corolla separate? Here a little doubt may arise; but suppose we answer —*Yes*. Then our plant will be found somewhere in the POLYPETALOUS DIVISION. Proceeding with the enquiries suggested under this heading:—Are the stamens more than twice as many as the petals? We find that they are not.

Turn, then, to the heading marked B, page xv, "Stamens not more than twice as many as the petals." Under this we find two subordinate headings, designated by asterisks * and **. The first of these is not applicable to our plant. Under the second, marked thus **, we find two minor headings, designated by daggers,+—and+—+—. The first of these, "*Corolla irregular*," is clearly the one we want. We have now, therefore, five Families to select from. We cannot choose any one of the first four, because our plant has ten stamens, but the characters of the fifth are precisely the characters exhibited by Clover. Our Clover, therefore, belongs to the Order LEGUMINOSÆ. Turning to page 33, and running through the "Synopsis of the Genera" as there given, we observe that No. 2, TRIFOLIUM, is the only Genus in which the flowers are in *heads*. Clover answers the description in the other respects also —viz. : "leaves of three leaflets," and "stamens diadelphous." The only question then remaining is, which Species of TRIFOLIUM have we in hand ? Turning to page 34, we find we have six Species to choose from. No. 2, TRIFOLIUM *pratense*, is the only one of them with *purplish flowers*. TRIFOLIUM *pratense* must, consequently, be the botanical name we are looking for.

Possibly the observer may decide that the parts of the corolla are not separate from each other, because in some instances it is really a doubtful question. He must then turn to page xvii, and under II. GAMOPETALOUS DIVISION, he must pursue his inquiries as before. Is the calyx superior ? Plainly not. Proceed then to the heading B, "Calyx inferior." Are the stamens more than the lobes of the corolla ? Yes. Then the choice of the six Orders in the Section marked * is easily made as before, and the plant is referred to LEGUMINOSÆ.

Now let us take the Strawberry. As with Clover, we decide without difficulty that the plant is a DICOTYLEDON. The carpels are separate, and produce achenes in fruit ; the plant must, therefore, be an ANGIOSPERM. And there is no doubt that it is Polypetalous. As the stamens are very numerous it must come under the section marked A. Under this section we have three subordinate headings, marked by one, two, and three asterisks, respectively. The stamens are clearly inserted on the calyx, and so our plant must be found under the heading marked **. Without hesitation, we refer it to the Order ROSACEÆ. Turning to page 38, we find fourteen Genera to select from. A very little consideration will show us that No. 8, FRAGARIA, is the Genus we must fix upon. Referring to page 43, we have to choose between two species, *Virginiana* and *vesca*, and the choice is found to depend upon such obvious characters as to furnish no difficulty.

The leaves of Blue Flag are straight-veined; the parts of the flower, also, are in threes. We therefore decide that the plant is Monocotyledonous, and on turning to page xxii, we find three Divisions of Monocotyledons. The Flag clearly belongs to the PETALOIDEOUS DIVISION. Then, is the perianth superior or inferior? Clearly the former. Next, are the flowers diœcious or perfect? Clearly perfect. And as the flower has three stamens, it must belong to the Order IRIDACEÆ, described on page 155. The Genus is at once seen to be IRIS, and the Species is determined without difficulty.

The Cat-tail Flag is also manifestly Monocotyledonous, from the veining of the leaves. But it is not Petaloideous. The flowers are collected on a more or less fleshy axis at the top of a scape. It therefore belongs to the SPADICEOUS DIVISION, in which there are four Orders. The only practical question is, whether our plant belongs to ARACEÆ or TYPHACEÆ. On the whole, we choose the latter, and find our decision confirmed on reading the fuller account of the two Orders on pages 143 and 144. The Genus is immediately seen to be TYPHA, and the Species *latifolia*.

These examples need not be extended here; but the beginner is recommended to run down, in the same manner, a few plants whose names he already knows. If successful in these attempts, he will naturally acquire confidence in his determinations of plants previously unknown to him.

KEY TO THE FAMILIES OR ORDERS

INCLUDED IN THIS WORK.

SERIES I. PHANEROGAMS.

Plants producing true flowers and seeds.

CLASS I. DICOTYLEDONS.

Distinguished ordinarily by having net-veined leaves, and the parts of the flowers in fours or fives, very rarely in sixes. Wood growing in rings, and surrounded by a true bark. Cotyledons of the embryo mostly two.

SUB-CLASS I. ANGIOSPERMS.

Seeds enclosed in an ovary.

I. POLYPETALOUS DIVISION.

Two distinct sets of Floral Envelopes. Parts of the corolla separate from each other.

● — A. Stamens more than twice as many as the petals.

— * *Stamens hypogynous (inserted on the receptacle).*

+ *Pistil apocarpous (carpels separate from each other).*

RANUNCULACEÆ —Herbs. Leaves generally decompound or much dissected.................................... 2
ANONACEÆ.—Small trees. Leaves entire. Petals 6, in 2 sets 7
MAGNOLIACEÆ—Trees. Leaves truncate. Fruit resembling a cone 6
MENISPERMACEÆ.—Woody twiners. Flowers diœcious. Leaves peltate near the edge................... 7

Brasenia, in
NYMPHÆACEÆ.—Aquatic. Leaves oval, peltate; the petiole attached to the centre..................... 9

MALVACEÆ.—Stamens monadelphous. Calyx persistent. Ovaries in a ring............................. 24

Podophyllum, in

BERBERIDACEÆ.—Calyx fugacious. Leaves large, peltate, deeply lobed. Fruit a large fleshy berry, 1-celled. 8

++ *Pistil syncarpous. (Stigmas, styles, placentæ, or cells, more than one.)*

Actæa, in

RANUNCULACEÆ, might be looked for here. Fruit a many-seeded berry. Leaves compound.......... 2

NYMPHÆACEÆ.—Aquatics. Leaves floating, large, deeply cordate..... 9

SARRACENIACEÆ.—Bog-plants. Leaves pitcher-shaped.... 10

PAPAVERACEÆ.—Juice red or yellow. Sepals 2, caducous. 10

CAPPARIDACEÆ.—Corolla cruciform, but pod 1-celled. Leaves of 3 leaflets.. 16

HYPERICACEÆ. — Leaves transparent-dotted. Stamens usually in 3, but sometimes in 5, clusters....:.... 19

CISTACEÆ.—Sepals 5, very unequal, or only 3. Ovary 1-celled, with 3 parietal placentæ................ 18

MALVACEÆ.—Stamens monadelphous, connected with the bottom of the petals. Calyx persistent. Ovaries in a ring.. 24

TILIACEÆ.—Trees. Flowers yellowish, in small hanging cymes, the peduncle with a leaf-like bract attached...................................... 25

* * *Stamens perigynous (inserted on the calyx).*

Portulaca, in

PORTULACACEÆ.—Low herbs, with fleshy leaves. Sepals 2, adhering to the ovary beneath. Pod opening by a lid............................... 23*

ROSACEÆ.—Leaves alternate, with stipules. Fruit apocarpous, or a drupe, or a pome................ 38

* * * *Stamens epigynous (attached to the ovary).*

Nymphæa, in

NYMPHÆACEÆ.—Aquatic. Leaves floating. Flowers white, large, with numerous petals gradually passing into stamens........................ 9

B. Stamens not more than twice as many as the petals.

* *Stamens just as many as the petals, and one stamen in front of each petal.*

BERBERIDACEÆ.—Herbs (with us). Anthers opening by uplifting valves.. 8
PORTULACACEÆ.—Sepals 2. Styles 3-cleft. Leaves 2, fleshy.. 23
VITACEÆ.—Shrubs, climbing by tendrils. Calyx minute. 29
RHAMNACEÆ.—Shrubs, not climbing 29

Lysimachia, in

PRIMULACEÆ, is occasionally polypetalous. Flowers yellow, in axillary spikes; the petals sprinkled with purplish dots.. 91

* * *Stamens either just as many as the petals and alternate with them, or not of exactly the same number.*

+*Corolla irregular.*

FUMARIACEÆ.—Corolla flattened and closed. Stamens 6. 11
VIOLACEÆ.—Corolla 1-spurred. Stamens 5. Pod with 3 rows of seeds on the walls........................... 17
BALSAMINACEÆ.—Corolla 1-spurred, the spur with a tail. Stamens 5. Pod bursting elastically............. 27
POLYGALACEÆ.—Lower petal keel-shaped, usually fringed at the top. Anthers 6 or 8, 1-celled, opening at the top. Pod 2-celled............................ 32
LEGUMINOSÆ.—Corolla mostly papilionaceous. Filaments often united. Ovary simple, with one parietal placenta. Leaves compound................... 33

++*Corolla regular, or nearly so.*

1. Calyx superior (*i.e.*, adherent to the ovary, wholly or partially).

(*a*) *Stamens perigynous* (*inserted on the calyx*).

Cratægus, in

ROSACEÆ.—Shrubs. Stamens occasionally from 5 to 10 only. Leaves alternate, with stipules. Fruit drupe-like, containing 1-5 bony nutlets..................... 38
SAXIFRAGACEÆ.—Leaves opposite or alternate, without stipules. Styles or stigmas 2; in one instance 4. Ovary 1-celled, with 2 or 3 parietal placentæ..... 46

KEY TO THE ORDERS.

HAMAMELACEÆ.—Shrubs. Stamens 8; styles 2. Flowers yellow, in autumn............................... 48
HALORAGEÆ.—Aquatics. Stamens 4 or 8. Styles or sessile stigmas 4................................. 49
ONAGRACEÆ.—Flowers symmetrical. Stamens 2, 4, or 8. Stigmas 2 or 4, or capitate.................... 49
MELASTOMACEÆ.—Anthers 1-celled, opening by a pore at the apex. Stamens 8. Style and stigma 1. Flowers purple..................................... 51
LYTHRACEÆ.—Calyx apparently adherent to, but really free from, the ovary. Stamens 10, in 2 sets. Leaves mostly whorled......................... 51
CUCURBITACEÆ.—Tendril-bearing herbs. Flowers monœcious... 52

(*b*) *Stamens epigynous (on the ovary, or on a disk which covers the ovary).*
Euonymus, in

CELASTRACEÆ.—Shrub, with 4-sided branchlets, not climbing. Leaves simple. Pods crimson when ripe. Calyx not minute............................... 30
UMBELLIFERÆ.—Flowers chiefly in compound umbels. Calyx very minute. Stamens 5. Styles 2. Fruit dry, 2-seeded.................................. 53
ARALIACEÆ.—Umbels not compound, but sometimes panicled. Stamens 5. Styles usually more than 2. Fruit berry-like................................. 56
CORNACEÆ.—Flowers in cymes or heads. Stamens 4. Style 1... 57

2. Calyx inferior (*i.e.*, free from the ovary).

(*a*) *Stamens hypogynous (on the receptacle).*

CRUCIFERÆ.—Petals 4. Stamens 6, tetradynamous. Pod 2-celled...................................... 12
CISTACEÆ.—Petals 3. Sepals 5, very unequal; or only 3. Pod partly 3-celled........................... 18
DROSERACEÆ.—Leaves radical, beset with reddish glandular hairs. Flowers in a 1-sided raceme......... 19
Elodes, in
HYPERICACEÆ.—Leaves with transparent dots. Stamens 9, in 3 clusters............................. 19

CARYOPHYLLACEÆ.—Styles 2-5. Ovules in the centre or bottom of the cell. Stem usually swollen at the joints. Leaves opposite........................ 21

LINACEÆ.—Stamens 5, united below. Pod 10-celled, 10-seeded... 25

GERANIACEÆ.—Stamens 5. Carpels 5,—they and the lower parts of the 5 styles attached to a long beak, and curling upwards in fruit........................ 26

OXALIDACEÆ.—Stamens 10. Pod 5-celled. Styles 5, distinct. Leaflets 3, obcordate, drooping at night-fall. 27

ERICACEÆ.—Anthers opening by pores at the top, or across the top. Leaves mostly evergreen, sometimes brown beneath ; but in some instances the plant is white or tawny....................................... 85

(*b*) *Stamens perigynous* (*plainly attached to the calyx*).

SAXIFRAGACEÆ.—Leaves opposite or alternate, without stipules. Styles or stigmas 2; in one instance 4. Carpels fewer than the petals.................. 46

CRASSULACEÆ.—Flowers *symmetrical*. Stamens 10 or 8. Leaves sometimes fleshy....................... 48

LYTHRACEÆ.—Stamens 10, in two sets. Calyx enclosing, but really free from, the ovary. Leaves mostly whorled... 51

(*c*) *Stamens attached to a fleshy disk in the bottom of the calyx-tube.*

ANACARDIACEÆ.—Trees, or shrubs, not prickly. Leaves compound. Stigmas 3. Fruit a 1-seeded drupelet. 28

CELASTRACEÆ.—Twining shrub. Leaves simple. Pods orange when ripe.. 30

SAPINDACEÆ.—Shrubs, or trees. Fruit 2-winged, and leaves palmately-veined. *Or*, Fruit an inflated 3-celled pod, and leaves of 3 leaflets. Styles 2 or 3...... 31

(*d*) *Stamens attached to the petals at their very base.*

Claytonia, in

PORTULACACEÆ.—Sepals 2. Leaves fleshy. Style 3-cleft. 23

AQUIFOLIACEÆ.—Shrubs, with small axillary flowers, having the parts in fours or sixes. Fruit a red berry-like drupe. Stigma sessile. Calyx minute....... 90

II. GAMOPETALOUS DIVISION.

Corolla with the petals united together, in however slight a degree.

A. Calyx superior (adherent to the ovary).

Stamens united by their anthers.

CUCURBITACEÆ.—Tendril-bearing herbs.................. 52
COMPOSITÆ.—Flowers in heads, surrounded by an involucre. 64
LOBELIACEÆ.—Flowers not in heads. Corolla split down
    one side........... 83

** *Stamens not united together in any way.*

+ *Stamens inserted on the corolla.*

DIPSACEÆ.—Flowers in heads, surrounded by an involucre.
    Plant prickly................................. 63
VALERIANACEÆ.—Flowers white, in clustered cymes. Sta-
    mens fewer than the lobes of the corolla.......... 63
RUBIACEÆ.—Leaves, when opposite, with stipules; when
    whorled, without stipules. Flowers, if in heads,
    without an involucre........................... 61
CAPRIFOLIACEÆ.—Leaves opposite, without stipules; but,
    in one genus, with appendages resembling stipules. 58

++ *Stamens not inserted on the corolla.*

CAMPANULACEÆ.—Herbs with milky juice. Stamens as
    many as the lobes of the corolla.................. 84
ERICACEÆ.—Chiefly shrubby plants or parasites. Stamens
    twice as many as the lobes of the corolla...... 85

B. Calyx inferior (free from the ovary).

* *Stamens more than the lobes of the corolla.*

LEGUMINOSÆ.—Ovary 1-celled, with 1 parietal placenta. Sta-
    mens mostly diadelphous........................ 33

Adlumia, in

FUMARIACEÆ.—Plant climbing. Corolla 2-spurred........ 11
MALVACEÆ.—Filaments monadelphous. Carpels in a ring. 24
ERICACEÆ.—Chiefly shrubby plants, with simple entire
    leaves. Stamens twice as many as the lobes of the
    corolla... 85
POLYGALACEÆ.—Anthers 6 or 8, 1-celled, opening at the
    top. Pod 2-celled. Flowers irregular; lower petal
    keel shaped, and usually fringed at the top........ 32
OXALIDACEÆ.—Stamens 10, 5 of them longer. Styles 5,
    distinct. Leaflets 3, obcordate, drooping at night-
    fall.. 27

KEY TO THE ORDERS. xix

* * *Stamens just as many as the lobes of the corolla, one in front of each lobe.*

PRIMULACEÆ.—Stamens on the corolla. Ovary 1-celled, with a free central placenta rising from the base. 91

* * * *Stamens just as many as the lobes of the corolla, inserted on its tube alternately with its lobes.*

          ÷ *Ovaries 2, separate.*

APOCYNACEÆ.—Plants with milky juice. Anthers converging round the stigmas, but not adherent to them. Filaments distinct............................... 114
ASCLEPIADACEÆ.—Plants with milky juice. Anthers adhering to the stigmas. Filaments monadelphous. Flowers in umbels.............................. 114

    ÷ ÷ *Ovary 4-lobed around the base of the style.*

Mentha, in

LABIATÆ.—Stamens 4. Leaves opposite, aromatic........ 100
BORRAGINACEÆ.—Stamens 5. Leaves alternate.......... 105

    ÷ ÷ ÷ *Ovary 1-celled ; the seeds on the walls.*

HYDROPHYLLACEÆ.—Stamens 5, usually exserted. Style 2-cleft. Leaves lobed and sometimes cut-toothed. 108
GENTIANACEÆ.—Leaves entire and opposite ; or (in Menyanthes) of 3 leaflets............................... 112

    — ÷ ÷ ÷ ÷ *Ovary with 2 or more cells.*

AQUIFOLIACEÆ.—Shrubs. Corolla almost polypetalous. Calyx minute. Fruit a red berry-like drupe. Parts of the flower chiefly in fours or sixes....... 90
PLANTAGINACEÆ.—Stamens 4. Pod 2-celled. Flowers in a close spike....................................... 91
~~~~~~~~~~~, in
SCROPHULARIACEÆ.—Corolla nearly regular. Flowers in a long terminal spike. Stamens 5 ; the filaments, or some of them, woolly........................ 94
POLEMONIACEÆ.—Style 3-cleft. Corolla salver-shaped, with a long tube. Pod 3-celled, few-seeded ; seeds small... 109
CONVOLVULACEÆ.—Style 2-cleft. Pod 2-celled, generally 4-seeded ; seeds large. Chiefly twining or trailing plants.. 109

SOLANACEÆ.—Style single. Pod or berry 2-celled, many-seeded. 119

* * * * *Stamens fewer than the lobes of the corolla; the corolla mostly irregular or 2-lipped.*

LABIATÆ.—Ovary 4-lobed around the base of the style. Stamens 4 and didynamous, or occasionally only 2 with anthers. Stem square.................. 100
VERBENACEÆ.—Ovary 4-celled, but not lobed; the style rising from the apex. *Or*, Ovary 1-celled and 1-seeded. Stamens didynamous.................. 99
LENTIBULACEÆ.—Aquatics. Stamens 2. Ovary 1-celled, with a free central placenta.................. 93
OROBANCHACEÆ.—Parasitic herbs, without green foliage. Ovary 1-celled, with many seeds on the walls. Stamens didynamous. 94
SCROPHULARIACEÆ.—Ovary 2-celled, with many seeds. Stamens didynamous, or only 2.................. 94

III. APETALOUS DIVISION.

Corolla (and sometimes calyx also) wanting.

A. Flowers not in catkins.

* *Calyx and corolla both wanting.*

SAURURACEÆ.—Flowers white, in a dense terminal spike, nodding at the end. Carpels 6 or 4, nearly separate. 124
CERATOPHYLLACEÆ.—Immersed aquatics, with whorled finely dissected leaves. Flowers monœcious...... 124

* * *Calyx superior (i.e., adherent to the ovary).*

SAXIFRAGACEÆ.—Small, smooth herbs, with inconspicuous greenish-yellow flowers. Stamens twice as many as the calyx-lobes, on a conspicuous disk......... 46
HALORAGEÆ.—Aquatics. Leaves finely dissected or linear. Stamens 1–8. Ovary 4-lobed or (Hippuris) 1-celled. 49
ONAGRACEÆ.—Herbs, in ditches. Stamens 4. Ovary 4-celled, 4-sided................................ 49
ARISTOLOCHIACEÆ. -- Calyx 3-lobed, dull purple inside. Ovary 6-celled.............................. 116
SANTALACEÆ. -Low plants with greenish-white flowers in terminal clusters. Calyx-tube prolonged, and forming a neck to the 1-celled nut-like fruit...... 124

ELÆAGNACEÆ.—Shrubs with scurfy leaves. Flowers diœcious. Calyx 4-parted, in the fertile flowers apparently adherent to the ovary, and becoming fleshy in fruit... 123

* * * *Calyx inferior (plainly free from the ovary).*

+ *Ovaries more than one and separate from each other.*

RANUNCULACEÆ.—Calyx present, coloured and petal-like. Achenes containing several seeds, or only one..... 2
RUTACEÆ.—Prickly shrubs, with compound transparent-dotted leaves, and diœcious flowers.............. 27

++ *Ovary only one, but with more than one cell.*

CRASSULACEÆ.—Herbs, in wet places. Pod 5-celled and 5-horned... 48
PHYTOLACCACEÆ.—Herbs. Ovary 10-celled and 10-seeded. 116
EUPHORBIACEÆ.—Herbs. Ovary 3-celled, 3-lobed, protruded on a long pedicel. Juice milky.......... 125
SAPINDACEÆ.—Trees. Ovary 2-celled and 2-lobed. Fruit two 1-seeded samaras joined together. Flowers polygamous... 31
RHAMNACEÆ.—Shrubs. Ovary 3-celled and 3-seeded; forming a berry.. 29
FICOIDEÆ.—Prostrate herbs with whorled leaves. Ovary 3-celled, many-seeded 52
URTICACEÆ.—Trees. Leaves simple. Ovary 2-celled, but fruit a 1-seeded samara winged all round. Stigmas 2.. 127

+++ *Ovary only one, 1-celled and 1-seeded.*

POLYGONACEÆ.—Herbs. Stipules sheathing the stem at the nodes... 119
URTICACEÆ.—Herbs. Stigma 1. Flowers monœcious or diœcious, in spikes or racemes. No chaff-like bracts among the flowers. *Or*, Stigmas 2; leaves palmately-compound................................. 127
AMARANTACEÆ.—Herbs. Flowers greenish or reddish, in spikes, *with chaff-like bracts interspersed.* Stigmas 2. 118
CHENOPODIACEÆ.—Herbs. Flowers greenish, in spikes. *No chaff-like bracts.* Stigmas 2................ 116
OLEACEÆ.—Trees. *Leaves pinnately-compound.* Fruit a 1-seeded samara... 115

URTICACEÆ.—Trees. *Leaves simple.* Fruit a 1-seeded samara winged all round, or a drupe 127
LAURACEÆ.—Trees or shrubs. Flowers diœcious. Sepals 6, petal-like. Stamens 9, opening by uplifting valves. 122
THYMELEACEÆ.—Shrubs with leather-like bark, and jointed branchlets. Flowers perfect, preceding the leaves. Style thread-like..... 123

B. **Flowers in catkins.**

* *Sterile or staminate flowers only in catkins.*

JUGLANDACEÆ. —Trees with pinnate leaves. Fruit a nut with a husk.............................. 130
CUPULIFERÆ.—Trees with simple leaves. Fruit one or more nuts surrounded by an involucre which forms a scaly cup or bur...................... 131

* * *Both sterile and fertile flowers in catkins, or catkin-like heads.*

SALICACEÆ.—Shrubs or low trees Ovary 1-celled, many-seeded ; seeds tufted with down at one end....... 136
PLATANACEÆ.—Large trees. *Stipules sheathing the branchlets.* The flowers in heads....... 130
MYRICACEÆ.—Shrubs with resinous-dotted, usually fragrant, leaves. Fertile flowers one under each scale. Nutlets usually coated with waxy grains... 134
BETULACEÆ.—Trees or shrubs. Fertile flowers 2 or 3 under each scale of the catkin. Stigmas 2, long and slender............ 135

SUB-CLASS II. GYMNOSPERMS.

Ovules and seeds naked, on the inner face of an open scale ; or, in Taxus, without any scale, but surrounded by a ring-like disk which becomes red and berry-like in fruit.
CONIFERÆ.—Trees or shrubs, with resinous juice, and mostly awl-shaped or needle-shaped leaves. Fruit a cone, or occasionally berry-like...... 139

CLASS II. MONOCOTYLEDONS.

Distinguished ordinarily by having straight-veined leaves (though occasionally net-veined ones), and the parts of the flowers in threes, never in fives. Wood never forming rings, but interspersed in separate bundles throughout the stem. Cotyledon only 1.

I. SPADICEOUS DIVISION.

Flowers collected on a spadix, with or without a spathe or sheathing bract. Leaves sometimes net-veined.

ARACEÆ.—Herbs (either flag-like marsh-plants, or terrestrial,) with pungent juice, and simple or compound leaves, these sometimes net-veined. Spadix usually (but not always) accompanied by a spathe. Flowers either without a perianth of any kind, or with 4–6 sepals.................................. 143

TYPHACEÆ.—Aquatic or marsh plants, with linear straight-veined leaves erect or floating, and monœcious flowers. Heads of flowers cylindrical or globular, no spathe, and no floral envelopes. 144

LEMNACEÆ.—Small aquatics, freely floating about........ 144

NAIADACEÆ.—Immersed aquatics. Stems branching and leafy. Flowers perfect, in spikes, generally on the surface. 145

II. PETALOIDEOUS DIVISION.

Flowers not collected on a spadix, furnished with a corolla-like, or occasionally herbaceous, perianth.

A. Perianth superior (adherent to the ovary).

Flowers diœcious or polygamous, regular.

HYDROCHARIDACEÆ.—Aquatics. Pistillate flowers only above water; perianth of 6 pieces... 148

DIOSCOREACEÆ.—Twiners, from knotted rootstocks. Leaves heart-shaped, net-veined. Pod with 3 large wings. 157

* * *Flowers perfect.*

ORCHIDACEÆ.—Stamens 1 or 2, gynandrous. Flowers irregular... 149

IRIDACEÆ.—Stamens 3.................................... 155

AMARYLLIDACEÆ.—Stamens 6. Flowers on a scape from a bulb.. 156

B. Perianth inferior (free from the ovary).

ALISMACEÆ.—Pistil apocarpous; carpels in a ring or head, leaves with distinct petiole and blade............ 147

SMILACEÆ—Climbing plants, with alternate ribbed and net-veined petioled leaves. Flowers diœcious........ 157

Triglochin, in

ALISMACEÆ.—Rush-like marsh herbs. Flowers in a spike or raceme. Carpels when ripe splitting away from a persistent axis. 147
LILIACEÆ.—Perianth of similar divisions or lobes, mostly 6, but in one case 4. One stamen in front of each division, the stamens similar. 150

Trillium, in

LILIACEÆ.—Perianth of 3 green sepals and three coloured petals 158
PONTEDERIACEÆ.—Stamens 6, 3 long and 3 short. Perianth (blue or yellow) tubular, of 6 lobes. Aquatics.... 164
JUNCACEÆ.—Perianth glumaceous, of similar pieces. 162
ERIOCAULONACEÆ.—In shallow water. Flowers in a small woolly head, at the summit of a 7-angled scape. Leaves in a tuft at the base. 163

III. GLUMACEOUS DIVISION.

Flowers without a true perianth, but subtended by thin scales called glumes.

CYPERACE.Æ.—Sheaths of the leaves not split. 165
GRAMINEÆ.—Sheaths of the leaves split on the side away from the blade 168

SERIES II. CRYPTOGAMS.

Plants without stamens and pistils, reproducing themselves by spores instead of seeds.

CLASS III. PTERIDOPHYTES.

Stems containing vascular as well as cellular tissue.

FILICES.—Spores produced on the fronds. 174
EQUISETACEÆ.—Spores produced on the under side of the shield-shaped scales of a terminal spike or cone... 181
LYCOPODIACEÆ.—Spore-cases produced in the axils of the simple leaves or bracts. 182

THE COMMONLY OCCURRING

WILD PLANTS OF CANADA,

AND MORE ESPECIALLY THOSE OF ONTARIO.

SERIES I.

FLOWERING OR PHANEROG'AMOUS PLANTS.

Plants producing Flowers (that is to say, Stamens and Pistils, and usually Floral Envelopes of some kind), and Seeds containing an Embryo.

CLASS I. DICOTYLE'DONS.

(See Sections 78-81, Part I., for characters of Class.)

SUB-CLASS I. AN'GIOSPERMS.

Seeds enclosed in a pericarp.

I. POLYPET'ALOUS DIVISION.

Plants with flowers having both calyx and corolla, the latter consisting of petals entirely separate from each other. (In some genera and species, however, petals are absent.)

ORDER I. **RANUNCULA'CEÆ.** (CROWFOOT FAMILY.)

Herbs or woody climbers, with an acrid colourless juice. Parts of the flower separate from each other. Corolla sometimes wanting. Stamens numerous. Pistil (with one or two exceptions) apocarpous. Fruit an achene, follicle, or berry. Leaves exstipulate, with the blades usually dissected, and petioles spreading at the base.

Synopsis of the Genera.

1. **Clem'atis.** Real petals none or stamen-like. Coloured sepals 4 or more, valvate in the bud. Fruit an achene, with the long and feathery style attached. Leaves all opposite. Plant climbing by the bending of the petioles.
2. **Anemo'ne.** Petals none or stamen-like. Coloured (white) sepals imbricated in the bud. Achenes many, in a head, pointed or tailed, not ribbed. Stem-leaves opposite or whorled, *forming an involucre remote from the flower.*
3. **Hepat'ica.** Petals none. Coloured sepals 6-9, whitish or bluish. Achenes many, not ribbed. Leaves all radical. *An involucre of 3 leaves close to the flower,* and liable to be mistaken for a calyx.
4. **Thalic'trum.** Petals none. Coloured sepals 4 or more, greenish. Achenes several, angled or grooved. No involucre. Stem-leaves alternate, decompound. Flowers in panicles or corymbs, mostly diœcious.
5. **Ranun'culus.** Sepals 5, deciduous. Petals generally 5, each with a pit or little scale on the inside of the claw. Achenes many, in heads, short-pointed. Stem-leaves alternate. Flowers solitary or corymbed, mostly yellow, rarely white.
6. **Cal'tha.** Petals none. Sepals 5-9, yellow. Fruit a many-seeded follicle. Leaves large, glabrous, heart-shaped or kidney-shaped, mostly crenate. Stem hollow and furrowed.
7. **Cop'tis.** Sepals 5-7, white, deciduous. Petals 5-7, yellow, with slender claws, and somewhat tubular at the apex. Carpels 3-7, on slender stalks. Fruit a follicle. Flowers on naked scapes. Leaves radical, shining, divided into three wedge-shaped leaflets, sharply toothed. Root fibrous, golden yellow.
8. **Aquile'gia.** Sepals 5, coloured. Petals 5, *each a long hollow spur.* Carpels 5. Follicles erect, many-seeded. Flowers very showy, terminating the branches. Leaves decompound.
9. **Actæ'a.** Sepals 4-5, caducous. Petals 4-10, with slender claws. Stamens many, with long filaments. *Fruit a many-seeded berry.* Flowers in a short thick raceme. Leaves decompound, leaflets sharply toothed.
10. **Cimicif'uga.** Sepals 4-5, caducous. Petals several, small, two-horned at the apex. Carpels 1-8, becoming pods. Flowers in long plume-like racemes.

RANUNCULACEÆ.

11. **Hydras'tis.** Petals none. Flower solitary. Sepals 3, petal-like, greenish-white. Carpels 12 or more, forming a head of crimson 1-2-seeded berries in fruit. Stem low, from a knotted yellow root-stock. Leaves simple, lobed.

1. CLEM'ATIS, L. VIRGIN'S BOWER.

C. Virginia'na, L. (COMMON VIRGIN'S BOWER.) A woody-stemmed climber. Flowers in panicled clusters, often diœcious, white. Leaves of 3 ovate leaflets, which are cut or lobed. Feathery tails of the achenes very conspicuous in the autumn.—Along streams and in swamps.

2. ANEMO'NE, L. ANEM'ONE.

1. **A. cylin'drica, Gray.** (LONG-FRUITED A.) Carpels very numerous, in an oblong woolly head about an inch long. Peduncles 2-6, long, upright, leafless. Stem-leaves in a whorl, twice or thrice as many as the peduncles, *long-petioled*. Sepals 5, greenish-white. Plant about two feet high, clothed with silky hairs.—Dry woods.

2. **A. Virginia'na, L.** (VIRGINIAN A.) Very much like the last, but larger. Also, the central peduncle only is naked, the others having each a pair of leaves about the middle, from whose axils other peduncles occasionally spring. Sepals greenish. Head of carpels oval or oblong.—Dry rocky woods and river-banks.

3. **A. Pennsylva'nica, L.** (*A. dichotoma, L.,* in Macoun's Catalogue.) (PENNSYLVANIAN A.) Carpels fewer and the head not woolly, but pubescent and spherical. *Stem-leaves sessile*, primary ones 3 in a whorl, but only a pair of smaller ones on each side of the flowering branches. Radical leaves 5-7-parted. Sepals 5, obovate, large and white. Plant hairy, scarcely a foot high.—Low meadows.

4. **A. nemoro'sa, L.** (WOOD A. WIND-FLOWER.) Plant not more than six inches high, nearly smooth, one-flowered. Stem-leaves in a whorl of 3, long-petioled, 3-5-parted. Sepals 4-7, oval, white, or often purplish on the back.—Moist places.

3. HEPAT'ICA, Dill. LIVER-LEAF. HEPATICA.

1. **H. acutil'oba, DC.** (SHARP-LOBED H.) Leaves with 3 (sometimes 5) acute lobes, appearing after the flowers. Petioles silky-hairy.—Woods in spring.

2. **H. tril'oba,** Chaix. (ROUND-LOBED H.) Leaves with 3 rounded lobes; those of the involucre also obtuse.—Dry rich woods in spring.

(The two species just described are included under ANEMONE in Macoun's Catalogue, the first named being *A. acutiloba, Lawson,* and the second *A. Hepatica, L.*)

4. THALIC'TRUM, Tourn. MEADOW-RUE.

1. **T. anemonoi'des,** Michx. (RUE-ANEMONE.) Stem low. Stem-leaves all in a whorl at the top. Roots tuberous. *Flowers several in an umbel,* by which character this plant is easily distinguished from Wood Anemone, which it otherwise resembles.—South-westward, in spring.

2. **T. dioi'cum,** L. (EARLY M.) Stem smooth, pale and glaucous, 1-2 feet high. *Flowers diœcious,* in ample panicles, purplish or greenish; the yellow anthers drooping and very conspicuous. Leaves alternate, decompound; leaflets with 5-7 rounded lobes.—Woods.

3. **T. Cornu'ti,** L. (TALL M.) Stem smooth or nearly so, 2-6 feet high. *Leaves sessile;* leaflets very much like No. 2. Flowers white, in compound panicles; *anthers not drooping,* filaments club-shaped.—Low wet meadows, and along streams.

5. RANUN'CULUS, L. CROWFOOT. BUTTERCUP.

1. **R. aquat'ilis,** L. (WHITE WATER-CROWFOOT.) *Foliage under water,* filiform. *Flowers white,* floating, each petal with a little pit on the inside of the claw. (Of this species there are several varieties; var. **trichopyllus,** Chaix, is the one here described, being the form commonly met with in Ontario and the eastern provinces.)

2. **R. multif'idus,** Pursh. (YELLOW WATER-CROWFOOT.) Like No. 1, but larger, and with *yellow flowers.*—Ponds and ditches.

3. **R. Flam'mula,** L., var. **reptans,** Meyer. (CREEPING SPEARWORT.) Stem reclining, rooting at the joints, *only 3-6 inches long.* Leaves linear, entire, remote. Flowers yellow, ⅓ of an inch broad.—Sandy and gravelly shores of ponds and rivers.

4. **R. aborti'vus,** L. (SMALL-FLOWERED C.) Petals shorter than the reflexed calyx. Stem erect, *very smooth*, slender. Radical leaves roundish, crenate, petiolate; stem-leaves 3-5-parted, sessile. Carpels in a globular head, each with a minute curved beak.—Shady hill-sides and wet pastures.

5. **R. scelera'tus,** L. (CURSED C.) Petals about the same length as the calyx. Stem thick, hollow, *smooth*. Radical leaves 3-lobed; stem-leaves 3-parted, uppermost almost sessile. *Head of carpels oblong.*—Wet ditches.

6. **R. recurva'tus,** Poir. (HOOKED C.) Petals shorter than the reflexed calyx. *Stem hirsute*, with stiff spreading hairs. Radical and cauline leaves about alike, long-petioled. Head of carpels globular, *each with a long recurved beak.*—Woods.

7. **R. Pennsylva'nicus,** L. (BRISTLY C.) Petals not longer than the calyx. *Stem hirsute.* Leaves ternately divided, divisions of the leaves stalked, unequally 3-cleft. *Head of carpels oblong, with straight beaks,* and so easily distinguished from No. 6.—Wet places.

8. **R. re'pens,** L. (CREEPING C.) Petals much longer than the calyx. Early-flowering stems ascending, *putting forth long runners during the summer.* Leaves ternate, divisions generally stalked, petioles hairy. Peduncles furrowed.—Wet places.

9. **R. bulbo'sus,** L. (BULBOUS C. BUTTERCUP.) Petals much longer than the calyx. Stem erect, *from a bulb-like base.* Flowers an inch broad, on *furrowed peduncles.*—Pastures. Rather rare.

10. **R. a'cris,** L. (TALL C. BUTTERCUP.) Much taller than No. 9. Petals much longer than the calyx. Stem upright, *no bulb at the base.* Peduncles *not furrowed.*

11. **R. fascicula'ris,** Muhl. (EARLY C.) Petals much longer than the calyx. Plant 5-9 inches high, erect, pubescent with silky hairs. Radical leaves appearing pinnate, the terminal division long-stalked, the lateral ones sessile. *Root a bundle of thickened fleshy fibres.*—Rocky woods and fields in spring.

6. **CAL'THA,** L. MARSH-MARIGOLD.

C. palustris, L. (MARSH-MARIGOLD.) Stem about a foot high, hollow, round, forking, very glabrous. Flowers golden yellow, 1-1½ inches broad.—Swamps and wet meadows. A very conspicuous plant in early spring.

7. COP'TIS, Salisb. GOLDTHREAD.

C. trifolia, Salisb. (THREE-LEAVED GOLDTHREAD.) Low and stemless. Scapes 1-flowered, with a single bract above the middle. Petals much smaller than the sepals.—On logs and about stumps in cedar-swamps.

8. AQUILE'GIA. Tourn. COLUMBINE.

A. Canadensis, L. (WILD COLUMBINE.) Stem branching, a foot or more in height, smooth. Leaves decompound; leaflets in threes. Flowers nodding, scarlet outside, yellow within. — Rocky woods and thickets.

9. ACTÆA, L. BANEBERRY.

1. **A. spica'ta**, L., var. **rubra**, Michx. (RED B.) *Raceme short*, breadth and length being about the same. Pedicels slender. *Berries red.*—Rich woods.

2. **A. alba**, Bigel. (WHITE B.) *Raceme longer than broad.* Pedicels thickened in fruit, cherry-coloured. *Berries white.*— Same localities as No. 1.

10. CIMICIF'UGA, L. BUGBANE.

C. racemo'sa, Ell. (BLACK SNAKEROOT.) Stem 3-6 feet high. Resembling a tall Actæa, but easily distinguished by its plume-like raceme of white flowers.—Along Lake Erie.

11. HYDRAS'TIS, L. ORANGEROOT. YELLOW PUCCOON.

H. Canadensis, L. A low plant, bearing a single radical leaf, and a pair of cauline ones near the summit of the simple stem. Leaves rounded, cordate, 5 7-lobed, very large when fully grown.—Wet meadows, in early summer, south-westward.

ORDER II. **MAGNOLIA'CEÆ.** (MAGNOLIA FAMILY.)

Trees or shrubs, with alternate entire or lobed (not serrate) leaves. Sepals 3, coloured, deciduous. Petals 6 9, deciduous. Stamens hypogynous, indefinite, separate; anthers adnate. Carpels numerous, in many rows on an elongated receptacle. Fruit resembling a cone.

1. LIRIODEN'DRON, L. TULIP-TREE.

The only Canadian species is

L. Tulipif'era, L. A large and stately tree, growing to a great height in many parts of the western peninsula of Ontario. Leaves large, truncate, or with a shallow notch at the end. Flowers large, showy, solitary; petals greenish-yellow, marked with orange. Fruit a dry cone, which, at maturity, separates into dry indehiscent fruits, like samaras.

ORDER III. **ANONA'CEÆ.** (CUSTARD-APPLE FAMILY.)

Trees or shrubs, with alternate and entire leaves, and solitary, axillary, perfect, hypogynous flowers. Sepals 3. Petals 6, in two sets, deciduous. Stamens numerous. Carpels few or many, fleshy in fruit.

1. ASIM'INA, Adans. NORTH AMERICAN PAPAW.

The only Canadian species is

A. tril'oba, Dunal. (COMMON PAPAW.) Found only in the Niagara peninsula. A small tree, not unlike a young beech in appearance, and forming thickets near Queenston Heights. Flowers purple, appearing before the leaves; the three outer petals much larger than the three inner ones. Fruit 2 to 3 inches long, edible.

ORDER IV. **MENISPERMA'CEÆ.** (MOONSEED FAMILY.)

Woody twiners, with peltate alternate leaves and small diœcious flowers. Sepals and petals yellowish-white, usually six of each, the petals in front of the sepals. Stamens numerous. Fruit a drupe, in appearance something like a small grape, with moon-shaped seeds.

1. MENISPER'MUM, L. MOONSEED.

The only Canadian species is

M. Canadense, L. (CANADIAN MOONSEED.) A twining plant, found, though not abundantly, in low grounds in rich woods. It may be pretty easily recognized by its usually 7-angled thin leaves, which are *peltate near the edge*. Fruit bluish-black.

ORDER V. **BERBERIDA'CEÆ**. (BARBERRY FAMILY.)

Herbs (or shrubs), with alternate, petiolate, divided leaves. Sepals and petals in fours, sixes, or eights (except in the genus Podophyllum), with the petals in front of the sepals. Stamens (except in Podophyllum) as many as the petals, one before each. Anthers usually opening by a valve at the top. Fruit berry-like, or a pod.

Synopsis of the Genera.

* *Petals and stamens 6.*
1. **Caulophyllum.** A purplish herb, flowering in early spring. Petals thick, much shorter than the sepals.

** *Petals 6-9. Stamens 8-18.*
2. **Podophyllum.** Petals 6-9. Stamens 12-18. Anthers *not* opening by uplifting valves. Fruit a large berry.
3. **Jeffersonia.** Petals and stamens mostly 8. Anthers opening by uplifting valves. Pod opening by a lid.

1. **CAULOPHYL'LUM**, Michx. BLUE COHOSH.

C. thalictroi'des, Michx. (BLUE COHOSH.) Plant 1-2 feet high, very glaucous and dull purple when young. Flowers yellowish-green, in a terminal small raceme, appearing in spring before the *decompound leaves* are developed. Sepals 6, with 3 little bractlets at their base. Petals 6, thick and somewhat kidney-shaped, much smaller than the sepals. Stamens 6, one before each petal. Ovary bursting soon after the flowering, and leaving the two drupe-like seeds naked on their rather thick stalks. Fruit bluish, ⅓ of an inch across.—Rich woods.

2. **PODOPHYL'LUM**, L. MAY-APPLE. MANDRAKE.

P. peltatum, L. Stem about 1 foot high. Flowerless stems with one large 7-9 lobed umbrella-like leaf, peltate in the centre; the flowering ones with two leaves, peltate near the edge, the flower nodding from the fork. Sepals 6, caducous. Petals 6-9, large and white. Stamens 12-18. Fruit large, oval, yellowish, not poisonous.—Found in patches in rich woods. The leaves and roots are poisonous.

3. **JEFFERSONIA**, Barton. TWIN-LEAF.

J. diphyl'la, Pers. A low plant, flowering in early spring; the solitary white flowers on naked scapes. Sepals 4, fugacious.

Petals 8. Stamens 8. Ovary pointed. Stigma 2-lobed. Pod pear-shaped, the top forming a lid. Leaves radical, long-petioled; the blades *divided into two leaflets with the outer margins lobed.*—Woods, chiefly in the western peninsula of Ontario.

ORDER VI. **NYMPHÆA'CEÆ.** (WATER-LILY FAMILY.)

Aquatic herbs with cordate or peltate, usually floating, leaves. Floating flowers on long immersed peduncles. Petals and stamens generally numerous.

Synopsis of the Genera.

1. **Brase'nia.** Sepals and petals each 3 (occasionally 4). Stamens 12-24. Leaves oval, peltate.
2. **Nymphæ'a.** Sepals 4-6. Petals numerous, *white*, imbricated in many rows, gradually passing into stamens, hypogynous or epigynous. Stamens epigynous. Stigmas radiating as in a Poppy-head.
3. **Nu'phar.** Sepals 5-6, *yellow*. Petals many, small and stamen-like. Stamens under the ovary.

1. BRASE'NIA, Schreber. WATER-SHIELD.

B. pelta'ta, Pursh. Stems and under surface of the leaves coated with jelly. Leaves oval, two inches across, peltate. Flowers small, purplish.—Ponds and slow-flowing streams.

2. NYMPHÆ'A, Tourn. WATER-LILY.

1. **N. odora'ta,** Ait. (SWEET-SCENTED WATER-LILY.) Leaves orbicular, cleft at the base to the petiole, 5-9 inches wide, often crimson underneath. *Flower very sweet-scented.* Ponds and slow streams.

Var. **minor,** Sims, has much smaller leaves and flowers, and the latter are often pink-tinted.

2. **N. tubero'sa,** Paine. (TUBER-BEARING W.) Leaves larger and more prominently ribbed than in No. 1, reniform-orbicular, green on both sides. Flower not at all, or only slightly, sweet-scented. Root-stocks producing tubers, which come off spontaneously.—Mostly in slow waters opening into Lake Ontario.

3. NUPHAR, Smith. YELLOW POND-LILY.

1. **N. ad'vena,** Ait. (COMMON Y. P.) Leaves floating, or emersed and erect, thickish, roundish or oblong, cordate. *Sepals* 6.—Stagnant water.

2. **N. lu'teum,** Smith. (SMALL Y. P.) Floating leaves usually not more than two inches across, the sinus very narrow or closed. Flowers hardly an inch across. *Sepals 5.*—Northward, in slow waters.

ORDER VII. **SARRACENIA'CEÆ.** (PITCHER-PLANT F.)

Bog-plants, easily distinguished by their pitcher-shaped leaves, all radical.

1. **SARRACE'NIA,** Tourn. SIDE-SADDLE FLOWER.

S. purpu'rea, L. (PURPLE S. HUNTSMAN'S CUP.) Hollow leaves with a wing on one side, purple-veined, curved, with the hood erect and open. Sepals 5, coloured, with 3 small bractlets at the base. Petals 5, fiddle-shaped, curved over the centre of the flower, deep purple. Ovary 5-celled, globose, the short style expanding above into a 5-angled umbrella, with a hooked stigma at each angle. Flowers on naked scapes, nodding.—Bogs.

ORDER VIII. **PAPAVERA'CEÆ.** (POPPY FAMILY.)

Herbs, with coloured juice and alternate leaves without stipules. Flowers polyandrous, hypogynous. Sepals 2, caducous. Petals 4-12. Stamens numerous, anthers introrse. Fruit a 1-celled pod, with numerous seeds.

1. **CHELIDO'NIUM,** L. CELANDINE.

C. majus, L. Petals 4, deciduous, crumpled in the bud. *Juice of the plant yellow.* Flower-buds nodding. Flowers small, yellow, in a kind of umbel. Fruit a smooth 1-celled slender pod, from which the two valves fall away, leaving the parietal placentas as a slender framework, with the seeds attached.—Waste places.

2. **SANGUINA'RIA,** Dill. BLOOD-ROOT.

S. Canadensis, L. Petals 8-12, not crumpled in the bud. Flower-buds not nodding. A stemless plant, with a thick rhizome which emits a *red juice* when cut, and sends up in early spring a single rounded, 5-7-lobed, thickish leaf, and a 1-flowered scape. *Flowers white.*—Rich woods.

ORDER IX. **FUMARIA'CEÆ.** (FUMITORY FAMILY.)

Smooth herbs, with brittle stems, watery juice, dissected leaves and irregular flowers. Sepals 2, very small. Corolla flattened and closed, of 4 petals, the two inner united by their tips over the anthers of the 6 stamens. Stamens in two sets of 3 each; filaments often united; the middle anther of each set 2-celled, the others 1-celled. Fruit a 1-celled pod.

Synopsis of the Genera.

1. **Adlu'mia.** Corolla 2-spurred. Petals all permanently united. *Plant climbing.*
2. **Dicen'tra.** Corolla 2-spurred. Petals slightly united, easily separated. Not climbing.
3. **Coryd'alis.** Corolla 1-spurred. Fruit a slender pod, many-seeded.

1. ADLU'MIA, Raf. CLIMBING FUMITORY.

A. cirrho'sa, Raf. A smooth vine, climbing by the petioles of its decompound leaves. Flowers in axillary pendulous clusters, pale pink.—Low and shady grounds.

2. DICEN'TRA, Bork. DUTCHMAN'S BREECHES.

1. **D. Cucullaria**, DC. (DUTCHMAN'S BREECHES.) Leaves all radical, multifid; these and the slender scape rising from a bulb-like rhizome of coarse grains. Flowers several in a raceme, whitish, spurs *divergent, elongated, acute, straight.*—Rich woods.

2. **D. Canadensis**, DC. (SQUIRREL CORN.) Underground shoots bearing small yellow tubers, something like grains of corn. Leaves very much as in No. 1. Corolla merely *heart-shaped; spurs very short and rounded.* Flowers greenish-white, fragrant. —Rich woods.

3. CORYD'ALIS, Vent. CORYDALIS.

1. **C. au'rea**, Willd. (GOLDEN CORYDALIS.) *Stems low and spreading.* Leaves dissected. *Flowers in simple racemes, golden yellow.* Pods pendulous.—Rocky river-margins and burnt woods.

2. **C. glauca**, Pursh. (PALE CORYDALIS.) *Stems upright,* 1-4 feet high. *Flowers in compound racemes, purplish tipped with yellow.* Pods erect.—Rocky woods.

ORDER X. **CRUCIF'ERÆ.** (CRESS FAMILY.)

Herbs with a pungent watery juice, alternate leaves without stipules, and regular hypogynous flowers in racemes or corymbs. Pedicels without bractlets. Sepals 4, deciduous. Petals 4, forming a cross-shaped corolla. Stamens 6, two of them shorter. Fruit a silique, or silicle. (See Chap. IV., Part I., for dissection of typical flower.) The genera are distinguished by the pods and seeds, the flowers in all cases being much alike. The seeds are exalbuminous, consisting entirely of the embryo, which is folded up in a variety of ways. The radicle may be bent so as to lie against the *edge* of the cotyledons, and the seed when cut through crosswise shows this section : o=o ; the cotyledons are then said to be *accumbent*. Or the radicle may be folded against the *back* of the cotyledon, showing this cross-section : ⊖, in which case the cotyledons are said to be *incumbent;* and if, besides, being incumbent, the cotyledons are doubled round the radicle, thus : ⟆, they are then *conduplicate*.

Synopsis of the Genera.

* Pod a silique (much longer than broad).

1. **Nastur'tium.** Flowers white or yellow. Pod terete, oblong-linear or ellipsoid. Seeds in two rows in each cell, globular, without a wing. Cotyledons accumbent.
2. **Denta'ria.** Flowers white or pale purple. Pod lanceolate, flat. Seeds wingless, on broad seed-stalks. *Stem-leaves 2 or 3 in a whorl;* stem naked below. Root-stock toothed or tuberous. Cotyledons accumbent.
3. **Cardam'ine.** Flowers white or rose-coloured. Pod linear or lanceolate, flat. Seeds wingless, on slender seed-stalks. Stem leafy below. Cotyledons accumbent.
4. **Ar'abis.** Flowers white or whitish. Pod linear or *elongated*, flattened, *the valves usually with a distinct mid-rib*. Stem leafy. Cotyledons accumbent.
5. **Erys'imum.** Flowers yellow. Pod linear, distinctly 4-sided. *Pedicels of the pods diverging from the stem.* Leaves simple. Cotyledons incumbent.
6. **Sisym'brium.** Flowers yellow. Pods awl-shaped, or 4-6-sided, close pressed to the stem, the valves 1-3-nerved. *Pods sessile* or nearly so. Leaves runcinate. Cotyledons incumbent.
7. **Bras'sica.** Flowers yellow. Pod linear or oblong, nearly terete, or 4-sided, *with a distinct beak extending beyond the end of the valves*. Cotyledons conduplicate.

* * *Pod a silicle (comparatively short).*
← *Silicle compressed parallel with the broad partition, or globular.*
8. **Draba.** Flowers white. Pod flat, *twisted when ripe, many-seeded.* Cotyledons accumbent.
9. **Alys'sum.** Flowers pale yellow or white. Pod orbicular, flat, 2–4-seeded.
10. **Camel'ina.** Flowers yellow. Pod pear-shaped, pointed, valves 1-nerved. Cotyledons incumbent.
— ← *Silicle compressed contrary to the narrow partition.*
11. **Capsel'la.** Flowers white. Pod obcordate-triangular, valves boat-shaped, *wingless. Seeds numerous.* Cotyledons incumbent.
12. **Thlas'pi.** Flowers white. Pod obovate or obcordate, *winged.* Seeds several. Cotyledons accumbent.
13. **Lepid'ium.** Flowers white or whitish. Pod roundish, very flat, the valves boat-shaped and *winged. Seeds solitary.*
← ← ← *Silicle fleshy, jointed.*
14. **Caki'le.** Flowers purplish. Pod 2-jointed, fleshy. Leaves fleshy. Cotyledons accumbent.

1. NASTUR'TIUM. R. Br. WATER-CRESS.

1. **N. officina'le,** R. Br. (WATER CRESS.) Flowers white. Stem spreading and rooting. Leaves pinnate; leaflets 3–11, roundish or oblong, nearly entire. Pods oblong-linear.—Ditches and streamlets.

2. **N. palustre,** DC. (MARSH CRESS.) Flowers yellow. Stem erect. Leaves pinnately parted, the lobes cut-toothed. Pods ovoid.—Wet places.

3. **N. lacus'tre,** Gray. (LAKE CRESS.) Flowers white. An aquatic plant, with the submerged leaves finely dissected; the leaves out of the water oblong, and either entire, serrate, or pinnatifid. Pods ovoid, 1-celled.

4. **N. Armora'cia,** Fries. (HORSERADISH.) Has escaped from gardens in many places. Flowers white. Root-leaves very large, oblong, and generally crenate; stem-leaves lanceolate. Pods globular. Roots very large.

2. DENTA'RIA, L. TOOTHWORT. PEPPER-ROOT.

1. **D. diphyl'la,** L. (TWO-LEAVED T.) Flowers white. *Stem-leaves 2, nearly opposite,* ternately divided. Root-stock toothed, pleasantly pungent to the taste.—Rich woods.

2. **D. lacinia'ta,** Muhl. (LACINIATE T.) Flowers purplish *Stem-leaves 3 in a whorl.* Root-stock jointed, scarcely toothed. —Along streams.

3. **CARDAMINE**, L. BITTER CRESS.

1. **C. rhomboi'dea**, DC. (SPRING CRESS.) Flowers white or (in var. **purpurea**) rose-purple. *Stem tuberous at the base.* Lower leaves round-cordate; upper nearly lanceolate; all somewhat angled or toothed.—Wet meadows.

2. **C. pratensis**, L. (CUCKOO-FLOWER. LADIES' SMOCK.) Flowers white or rose-colour, showy. Stem from a short rootstock. Leaves pinnate, leaflets 7-15, those of the lower leaves rounded and stalked, entire or nearly so.—Bogs.

3. **C. hirsu'ta**, L. (SMALL BITTER CRESS.) Flowers white, small. Root fibrous. Leaves pinnate, leaflets 5-11, the terminal leaflets largest. Pods erect, slender.—Wet places.

4. **AR'ABIS**, L. ROCK CRESS.

* *Seeds in one row in each cell, nearly as broad as the partition.*

1. **A. lyra'ta**, L. (LOW R.) Flowers white, twice as long as the calyx. Radical leaves clustered, pinnatifid, the terminal lobe largest; stem-leaves scattered. Pods slender, erect, and spreading.—Rocky or sandy shores.

2. **A. hirsu'ta**, Scop. (HAIRY R.) Flowers greenish-white, small, slightly longer than the calyx. Stem-leaves many, rough, sagittate. *Pods erect, straight.* Stems 1-2 feet high, 2 or 3 from the same root.—Rocky shores and dry plains.

3. **A. læviga'ta**, DC. (SMOOTH R.) Flowers white, rather small. Leaves linear or lanceolate, entire or slightly toothed, sagittate, clasping. Pods long and narrow, *recurved-spreading.* Stem glaucous, 1-2 feet high.—Dry hill-sides. Easily recognized by the pods.

4. **A. Canadensis**, L. (SICKLE-POD.) Flowers whitish, with linear petals, about twice the length of the calyx. Stem-leaves pointed at both ends, downy. Pods 2-3 inches long, *scythe-shaped, hanging.* Stem 2-3 feet high. A striking plant when the pods are fully formed.—Dry woods and ravines.

** *Seeds in two distinct rows in each cell, narrower than the partition.*

5. **A. perfolia'ta**, Lam. (TOWER MUSTARD.) Flowers yellowish-white. Petals scarcely longer than the calyx. Stem 2-4 feet high, *glaucous*. Cauline leaves ovate-lanceolate or oblong.

clasping with sagittate base. Pods long and *very narrow, on erect pedicels.*—Meadows and old fields. Pretty easily recognized by its strict habit.

6. **A. Drummond'ii**, Gray. Flowers white or rose-coloured. Petals twice as long as the calyx. Stem 1-2 feet high, *smooth above.* Cauline leaves lanceolate or oblong-linear, with sagittate base. Pods long and flat; *the pedicels not so strictly erect* as in the last species.—Rocky banks of streams.

5. **ERYS'IMUM, L.** Treacle Mustard.

E. cheiranthoi'des, L. (Worm-seed Mustard.) Flowers yellow; inconspicuous. Leaves lanceolate, scarcely toothed, roughish with appressed pubescence. —Waste wet places.

6. **SISYM'BRIUM, L.** Hedge Mustard.

S. officina'le, Scop. (Hedge Mustard.) Flowers yellow, small. Leaves runcinate. Stem 1-2 feet high, with spreading branches.—A very common roadside weed.

7. **BRAS'SICA,** Tourn. Cabbage, Mustard, Etc.

1. **B. Sinapis'trum**, Bois. (Charlock.) Flowers bright yellow. Stem 1-2 feet high, branching, it and the leaves hairy. —Too common in our grain-fields.

2. **B. nigra.** (Black Mustard.) Flowers sulphur-yellow. Stem 3-6 feet high, round, smooth, and branching. Lower leaves lyrate.—Fields and waste places.

8. **DRABA, DC.** Whitlow-Grass.

D. arab'isans, Michx. Flowers white. Stem leafy, erectly branched, pubescent. Leaves lanceolate or linear, minutely dentate. Raceme short, erect. Pods half an inch long, twisted when ripe.—Rocky places.

9. **ALYS'SUM,** Tourn. Alyssum.

A. calyci'num, L. A dwarf hoary annual, with linear-spathulate leaves. Calyx persistent. Pod 4-seeded, sharp-edged.—Rather rare.

10. **CAMEL'INA,** Crantz. False Flax.

C. sati'va, Crantz. (Common F. Flax.) Flowers yellowish. Stem 1-2 feet high, straight, erect, branching. Leaves lanceolate, sagittate. Pods pear-shaped, large, margined.—In flax fields.

11. CAPSEL'LA, Vent. SHEPHERD'S PURSE.

C. Bursa-pasto'ris, Mœnch. Flowers small, white. Root-leaves clustered, pinnatifid; stem-leaves clasping, sagittate.—A very common weed.

12. THLASPI, Tourn. PENNYCRESS.

T. arvense, L. (FIELD PENNYCRESS.) A low smooth plant, with undivided radical leaves, and stem-leaves sagittate and clasping. Pods half an inch broad, deeply notched at the top.—Waste places.

13. LEPID'IUM, L. PEPPERGRASS.

1. **L. Virgin'icum,** L. (WILD P.) Flowers small; *petals present,* white. Stem 1-2 feet high. Leaves lanceolate, the upper linear or lanceolate and entire, the lower toothed or pinnatifid, tapering towards the base. Pods marginless or nearly so, oval or orbicular.—Along railways and roadsides.

2. **L. interme'dium,** Gray. Distinguished from No. 1 by having the cotyledons incumbent instead of accumbent, and the pods minutely winged at the top.—Dry sandy fields.

3. **L. rudera'le,** L. *Petals always absent.* More branched than the preceding.

4. **L. campestre,** L. Well distinguished from other species by its *sagittate, clasping* leaves. Pods ovate, winged.—Rather rare.

14. CAKILE, Tourn. SEA-ROCKET.

C. America'na, Nutt. (AMERICAN S.) Flowers purplish. Leaves obovate, fleshy, wavy-toothed. Pod fleshy, 2-jointed.—Seashore, and borders of the Great Lakes.

ORDER XI. **CAPPARIDA'CEÆ.** CAPER FAMILY.

Herbs (in Canada), with an acrid watery juice, and alternate palmately-compound leaves. Flowers cruciform. Stamens 8 or more. Pod like that of a crucifer, *but only 1-celled.*

VIOLACEÆ.

1. POLANIS'IA, Raf. POLANISIA.

The only species in Canada is

P. grave'olens, Raf. A strong-scented herb, with a viscid, hairy stem. Leaflets 3. Flowers in terminal racemes. Sepals 4. Petals 4, yellowish-white, narrowed below into long claws. Stamens 8-12, exserted. Pod glandular-pubescent, 2 inches long, linear.—Shore of Lake Ontario, Hamilton to Niagara.

ORDER XII. **VIOLA'CEÆ.** (VIOLET FAMILY.)

Herbs, with alternate stipulate leaves. Flowers irregular, the lower of the 5 petals being spurred. Sepals 5, persistent. Stamens 5, the anthers slightly united and surrounding the pistil. Fruit a 1-celled pod, splitting into 3 valves. Seeds in 3 rows on the walls of the ovary. The only genus represented in this country is

VI'OLA, L. VIOLET.

* *Stemless Violets; leaves and scapes all from root-stocks.*
← *Flowers white.*

1. **V. blanda**, Willd. (SWEET WHITE V.) Lower petal streaked with purple. Leaves round, heart-shaped or reniform. Petals beardless. Flower sweet-scented.—Swamps and wet meadows, in spring.

2. **V. renifo'lia**, Gray. (KIDNEY-LEAVED V.) Leaves *much larger and more pubescent* than those of the preceding.—Dry cedar swamps, and ravines in rich woods.

← ← *Flowers blue or purple.*

3. **V. Selkirk'ii**, Pursh. (GREAT-SPURRED V.) A small and delicate plant, distinguished from the two following species by the *slender* root-stock, and the *very large spur, thickened at the end.* The pale violet petals, also, are *beardless.*—Damp, shady places.

4. **V. cuculla'ta**, Ait. (COMMON BLUE VIOLET.) *Leaves on very long petioles,* cordate or reniform, the sides folded inwards when young. Lateral petals bearded. Spur short and thick. —Low grounds everywhere.

5. **V. sagitta'ta**, Ait. (ARROW-LEAVED V.) Smoothish. Leaves cordate, halberd-shaped, or sagittate, slightly toothed,

the first ones on short and margined petioles. Side-petals bearded. —Dry hill-sides and old pastures.

** *Leafy-stemmed Violets.*
← *Flowers yellow.*

6. **V. pubes'cens,** Ait. (DOWNY YELLOW V.) Plant downy. Leaves broadly cordate, coarsely serrate ; stipules large, dentate. Lower petals veined with purple. Spur very short.—Rich woods.

← ← *Flowers not yellow.*

7. **V. Canadensis,** L. (CANADA VIOLET.) Tall, often a foot high. Leaves large, cordate, serrate-pointed. Petals white inside, *purplish outside*. Spur very short.— Flowering all summer.

8. **V. cani'na,** L., var. **sylvestris,** Regel. (DOG V.) Low, spreading by runners. Leaves broadly cordate or reniform, *with fringed-toothed stipules*. Spur cylindrical, half as long as the petals, which are *pale purple.*—Wet places.

9. **V. stria'ta,** Ait. (PALE V.) Stem angular, 6-10 inches high. Leaves cordate, finely serrate ; stipules fringed-toothed. Spur thickish, much shorter than the *cream-coloured or white petals.—*Low grounds.

10. **V. rostra'ta,** Pursh. (LONG-SPURRED V.) Distinguished at once by its extremely long straight spur. Petals violet-coloured.

ORDER XIII. **CISTA'CEÆ.** (ROCK-ROSE FAMILY.)

Herbs or low shrubs, with simple entire leaves and regular polyandrous flowers. Calyx persistent, usually of 3 large and 2 smaller sepals. Petals 5 or 3, convolute in the bud. Stamens 3-20. Pod 1-celled, 3-valved. Seeds on 3 parietal projections.

Synopsis of the Genera.

1. **Helian'themum.** Petals 5, fugacious. Style none.
2. **Hudso'nia.** Petals 5, fugacious. Style long and slender.
3. **Lech'ea.** Petals 3, persistent. Style none.

1. **HELIAN'THEMUM,** Tourn. ROCK-ROSE.

H. Canadense, Michx. (FROST-WEED.) Flowers of 2 sorts, some solitary, with large yellow corolla and many stamens, the petals lasting but one day after the flower opens; others small, clustered in the axils of the leaves, and apetalous. Leaves lanceolate, downy beneath.—Sandy places.

2. HUDSO'NIA, L. HUDSONIA.

H. tomento'sa, Nutt. (DOWNY H.) *Hoary*. Leaves oval or narrowly oblong, short, close-pressed, or imbricated. Flowers small, *yellow, very numerous*.—A little heath-like shrub, on the shores of the Great Lakes, and the River St. Lawrence.

3. LECH'EA, L. PINWEED.

L. minor, Lam. (SMALLER P.) Flowers inconspicuous, purplish, loosely racemose, on distinct pedicels. Stem slender, rough with appressed scattered hairs. Leaves scattered, linear. Pods the size of a pin's head.—Dry soil.

ORDER XIV. DROSERA'CEÆ. (SUNDEW FAMILY.)

Low glandular-hairy marsh herbs, with circinate tufted radical leaves, and regular hypogynous flowers borne on a naked scape. Sepals, petals, and stamens, 5 each; anthers turned outwards. Styles 3–5, deeply 2-parted. Pod 1-celled, 3-valved. The only genus with us is

DROS'ERA, L. SUNDEW.

1. **D. rotundifo'lia**, L. (ROUND-LEAVED SUNDEW.) Flowers small, white, in a 1-sided raceme. Leaves orbicular, abruptly narrowed into the hairy petiole, clothed with reddish glandular hairs.—Bogs.

2. **D. longifo'lia**, L. (LONGER-LEAVED S.) has oblong-spathulate leaves gradually narrowed into erect naked petioles.—Bogs; not common.

ORDER XV. HYPERICA'CEÆ. (ST. JOHN'S WORT F.)

Herbs or shrubs, with *opposite entire dotted leaves*, and no stipules. Flowers regular, hypogynous, mostly yellow. Sepals 5, persistent. Petals 5, deciduous. Stamens mostly numerous, and *usually in 3 or more clusters*. Styles 3–5, sometimes united. Pod 1–5-celled. Seeds numerous.

COMMON CANADIAN WILD PLANTS.

Synopsis of the Genera.

1. Hyper'icum. Petals 5, *unequal-sided*, convolute in the bud, *yellow*.
2. Elo'des. Petals 5, *equal-sided*, imbricated in the bud, *purplish*.

1. HYPER‑ICUM, L. ST. JOHN'S WORT.

* *Pod 3-celled. Styles 3, separate. Petals with black dots.*

1. **H. perfora'tum**, L. (COMMON ST. JOHN'S WORT.) Stem much branched, *producing runners at the base*, slightly 2-edged. Leaves linear-oblong, *with transparent dots*, easily observed by holding the leaf up to the light. *Petals deep yellow.* Flowers in open leafy cymes.—Fields.

2. **H. corymbo'sum**, Muhl. (CORYMBED S.) Stem rounded, not so branching as No. 1. *Leaves with both black and transparent dots*, oblong, somewhat clasping. Flowers *small, pale yellow, crowded.*—Damp woods and wet places generally.

** *Pod 5-celled. Styles more or less united. Stamens very many, in 5 clusters, if clustered at all.*

3. **H. pyramida'tum**, Ait. (GREAT ST. JOHN'S WORT.) Stem 3-5 feet high. Leaves 2-3 inches long, somewhat clasping. *Flowers very large, the petals about an inch long*, and narrowly obovate. Stamens showy. Pod conical, large.—Along streams; not common.

4. **H. Kalmia'num**, L. (KALM'S S.) *Shrubby*, a foot or more in height; leaves linear-lanceolate, crowded, revolute on the margins, thickly punctate, and sessile. Flowers about 1 inch across, in clusters.—Niagara Falls and westward.

* * * *Pod 1-celled, purple.*

5. **H. ellip'ticum**, Hook. (ELLIPTICAL-LEAVED S.) Stem about 1 foot high, *not branched*. Leaves spreading, elliptical-oblong, obtuse, thin. Flowers rather few, showy, in a nearly naked cyme. Pod purple, ovoid, obtuse. Petals pale yellow.—Banks of streams, eastward.

6. **H. mu'tilum**, L. (SMALL S.) Stem slender, branching above, hardly a foot high. Leaves 5-nerved. Cymes leafy at the base. *Flowers small*, not ¼ of an inch across.—Low grounds.

7. **H. Canadense**, L. (CANADA S.) Stem upright, 6-15 inches high, with branches erect. Leaves linear or linear-lanceolate, 3-nerved at the base, the upper ones acute, sessile. Cymes

naked. Pod much longer than the calyx. Flowers small, deep yellow.—Wet, sandy places.

2. ELO'DES. Adans. MARSH ST. JOHN'S WORT.

E. Virgin'ica, Nutt. Stem smooth. Leaves oblong or oval, clasping, often purple-veined, obtuse, conspicuously dotted beneath. Flowers flesh-coloured in the axils, and at the summit of the stem. The whole plant is of a purplish hue.—Marshes.

ORDER XVI. **CARYOPHYLLA'CEÆ.** (PINK FAMILY.)

Herbs with opposite and entire leaves, *the stems swollen at the joints.* Flowers regular, with the parts mostly in fives, occasionally in fours. Stamens not more than twice as many as the petals. Styles 2-5, stigmatic along the inner side. Pod usually 1-celled, with the seeds attached to the base, or to a column which rises from the centre of the cell. (Part I., Fig. 194.)

Synopsis of the Genera.

* *Sepals united into a tube or cup. Petals and stamens borne on the stalk of the ovary · petals with long narrow claws.*
1. **Sapona'ria.** Calyx cylindrical. Styles 2.
2. **Sile'ne.** Calyx 5-toothed. Styles 3.
3. **Lych'nis.** Calyx 5-toothed. Styles 5.

* * *Sepals separate to the base or nearly so. Petals without claws, they and the stamens inserted at the base of the sessile ovary.*
4. **Arena'ria.** Petals not cleft at the apex. Styles usually 3. Pod splitting into 3 or 6 valves.
5. **Stella'ria.** Petals 2-cleft at the apex. Pod splitting to the base into twice as many valves as there are styles. Styles generally 3.
6. **Ceras'tium.** Petals 2-cleft, or notched. Styles 5. Pod opening at the apex by 10 teeth.

1. SAPONA'RIA, L. SOAPWORT.

S. officina'lis, L. (BOUNCING BET.) A stout plant, with rose-coloured or pinkish flowers clustered in corymbs. Leaves 3-5-ribbed, the lower ovate, upper lanceolate. Pod raised on a short stalk. Styles 2.—Old gardens and roadsides.

2. SILE'NE, L. CATCHFLY. CAMPION.

1. **S. infla'ta,** Smith. (BLADDER CAMPION.) Pale or glaucous, very smooth. Stem erect, a foot high. Leaves ovate-lanceolate. *Calyx much inflated, purple-veined.* Stamens and styles exserted. —Not common westward.

2. **S. antirrhi'na,** L. (SLEEPY C.) Stem slender, simple or slightly branching above, a portion of the upper internodes sticky. Leaves linear or lanceolate. Flowers small, pink or purplish, opening only for a short time in sunshine. Calyx ovoid, shining.—Dry soil.

3. **S. noctiflo'ra,** L. (NIGHT-FLOWERING CATCHFLY.) *Stems very sticky, pubescent.* Lower leaves spathulate, upper lanceolate. Flowers few, *peduncled.* Calyx-tube with awl-shaped teeth. Petals white or whitish, 2-parted. Opening only at night or in cloudy weather.—A very common weed in cultivated grouds.

4. **S. Virgin'ica,** L. (FIRE PINK.) Occurs in south-western Ontario, and may be recognized by its *crimson petals,* and bell-shaped calyx, nodding in fruit.

3. LYCHNIS, Tourn. COCKLE.

L. Githa'go, Lam. (CORN COCKLE.) Plant clothed with long soft appressed hairs. *Calyx-lobes extremely long,* very much like the upper leaves, surpassing the *purple petals.*—Wheat-fields.

4. ARENA'RIA, L. SANDWORT.

1. **A. serpyllifo'lia,** L. (THYME-LEAVED S.) Much branched, 2-6 inches high, roughish-pubescent. Leaves small, ovate, acute. Petals white, hardly as long as the sepals. Sepals pointed, 3-5-nerved. Pod pointed, 6-toothed.—Sandy fields.

2. **A. stricta,** Michx. (*A. Michauxii,* Hook., in Macoun's Catalogue.) Stems erect, or diffusely spreading from a small root. Leaves awl-shaped or bristle form, the upper ones reduced to 1-nerved bracts, crowded in the axils. Cyme diffuse, many-flowered. Sepals pointed, *3-ribbed,* half as long as the white petals.—Rocky fields.

3. **A. lateriflo'ra,** L. Stem erect, slender, minutely pubescent. Leaves oval or oblong, $\frac{1}{2}$-1 inch long. Peduncles usually three-flowered. Sepals obtuse. Petals white, large, twice as long as the sepals. Flower $\frac{1}{3}$ of an inch across when fully expanded.—Gravelly shores.

4. **A. peploi'des,** L., with *very fleshy stems and leaves,* the latter somewhat clasping, occurs eastward towards the sea-coast.

5. **STELLA'RIA**, L. Chickweed. Starwort.

1. **S. media**, Smith. (Common Chickweed.) Stems branching, decumbent, soft and brittle, *marked lengthwise with one or two pubescent lines.* Lower leaves on hairy petioles, *ovate.* Flowers small, white. *Petals shorter than the sepals.*—Extremely common in damp grounds and old gardens.

2. **S. longifo'lia**, Muhl. (Long-leaved Stitchwort.) Stems branching, very weak and brittle, supporting themselves on other plants. *Leaves linear.* Pedicels of the flowers long, slender, and spreading, reflexed. Petals white, longer than the 3-nerved sepals. Low grassy banks of streams.

6. **CERAS'TIUM**, L. Mouse-ear Chickweed.

1. **C. vulga'tum**, L. (Common M.) Stem ascending, hairy and somewhat clammy. *Leaves ovate or obovate, obtuse.* Flowers in *close clusters. Pedicels not longer than the sepals.* Petals shorter than the calyx.—Not common, sometimes confounded with No. 2.

2. **C. visco'sum**, L. (Larger M.) Stems hairy, viscid, spreading. *Leaves lanceolate-oblong, rather acute.* Flowers in loose cymes. *Pedicels longer than the sepals.* Petals equalling the calyx.—Fields and copses; common.

3. **C. arven'se**, L. (Field Chickweed.) Stem decumbent at the base, pubescent, slender, 4-8 inches high. Leaves linear, or linear-lanceolate, *often fascicled in the axils*, longer than the lower internodes. Petals obcordate, more than twice as long as the calyx. Pod scarcely longer than the calyx. Cyme few-flowered.

4. **C. nu'tans**, Raf. Stems very clammy-pubescent and branching diffusely. The loose and open cymes many-flowered. Leaves lance-oblong. *Pods nodding on the stalks, curved upwards, thrice the length of the calyx.*—In places where water lies in spring.

Order XVII. **PORTULACA'CEÆ**. (Purslane F.)

Herbs with fleshy entire exstipulate leaves, and regular hypogynous or perigynous flowers. Sepals 2. Petals 5. Stamens 5-20. Styles 3-8, united below. Pod 1-celled, few or many-seeded.

Synopsis of the Genera.

1. **Portula'ca.** Stamens 8-20. Pod opening by a lid (Fig. 207, Part 1.), many-seeded.
2. **Clayto'nia.** Stamens 5. Pod 3-valved, 3-6-seeded.

1. PORTULA'CA, Tourn. PURSLANE.

P. olera'cea, L. (COMMON PURSLANE.) A low fleshy herb, very smooth, with obovate or wedge-shaped leaves. Calyx 2-cleft, the sepals keeled. Petals yellow, fugacious.—A common pest in gardens.

2. CLAYTO'NIA, L. SPRING-BEAUTY.

1. **C. Virgin'ica,** L. Leaves *linear-lanceolate*, 3-6 inches long.
2. **C. Carolinia'na,** Michx. Leaves *ovate-lanceolate* or oblong, tapering at the base. In both species the corolla is rose-coloured, with dark veins. The stem springs from a small tuber, and bears two opposite leaves and a loose raceme of flowers.—Rich woods in early spring.

ORDER XVIII. MALVA'CEÆ. (MALLOW FAMILY.)

Herbs, with palmately-veined alternate stipulate leaves. Flowers regular. Calyx valvate. Corolla convolute in the bud. Sepals 5. united at the base. Petals 5, hypogynous. Stamens numerous, monadelphous, hypogynous; anthers 1-celled. Carpels united in a ring, separating after ripening. Seeds kidney-shaped.

Synopsis of the Genera.

1. **Malva.** Carpels without beaks, 1-seeded. A circle of 3 bractlets at the base of the calyx.
2. **Abu'tilon.** Carpels 2-beaked, 1-6-seeded. No circle of bractlets.

1. MALVA, L. MALLOW.

1. **M. rotundifo'lia,** L. (ROUND-LEAVED MALLOW.) Stems several, procumbent, from a stout tap-root. Leaves long-petioled, round-heart-shaped, crenate, crenately-lobed. Petals obcordate, whitish, streaked with purple, twice as long as the sepals.—Waysides and cultivated grounds.

2. **M. sylves'tris,** L. (HIGH M.) *Stem erect*, 2 feet high. *Leaves sharply 5-7-lobed. Petals purple*, 3 times as long as the sepals.—Near dwellings.

TILIACEÆ, LINACEÆ.

3. M. moscha'ta, L. (MUSK M.) Stem erect, 1 foot high. *Stem-leaves 5-parted, the divisions cleft.* Flowers large and handsome, rose-coloured or white, on short peduncles, crowded on the stem and branches.—Roadsides near gardens.

2. ABU'TILON, Tourn. INDIAN MALLOW.

A. Avicen'næ, Gærtn. (VELVET-LEAF.) Stem 2–5 feet high, branching. Leaves velvety, round-cordate, long-pointed. Corolla yellow.—Near gardens; not common.

ORDER XIX. **TILIA'CEÆ.** (LINDEN FAMILY.)

Trees with fibrous bark, soft and white wood, and heart-shaped and serrate leaves, with deciduous stipules. Flowers in small cymes hanging on an axillary peduncle, to which is attached a leaf-like bract. Sepals deciduous. The only Canadian genus is

TILIA, L. BASSWOOD. WHITEWOOD.

T. America'na, L. (BASSWOOD.) A fine tree, in rich woods. Flowers yellow or cream-coloured, very fragrant. Leaves smooth and green on both sides, obliquely cordate or truncate at the base, sharply serrate. Sepals 5. Petals 5. Fruit a globular nut, 1-celled, 1–2-seeded.

ORDER XX. **LINA'CEÆ.** (FLAX FAMILY.)

Herbs with entire exstipulate leaves, and regular hypogynous flowers. Sepals, petals, stamens, and styles, 5 each. Filaments united at the base. Pod 10-celled, 10-seeded. Our only genus is

LINUM, L. FLAX.

1. **L. Virginia'num,** L. (VIRGINIA F.) *Flowers yellow, small* (⅓ of an inch long), scattered. Stem erect, it and the spreading branches terete. Leaves lanceolate and acute, the lower obtuse and opposite.—Dry soil.

2. **L. stria'tum,** Walt., has the branches *wing-angled*, broader leaves and more crowded flowers than No 1. The whole plant is stouter.

3. **L. usitatis'simum,** L. (COMMON F.) *Flowers blue.* Leaves alternate, linear-lanceolate, acute, 3-veined.—Cultivated grounds.

ORDER XXI. **GERANIA'CEÆ.** (GERANIUM FAMILY.)

Strong-scented herbs with pentamerous and symmetrical flowers, the filaments usually united at the base, and 5 glands on the receptacle alternate with the petals. Style 5-cleft. Carpels 5, each 2-ovuled (but 1-seeded), they and the lower part of the long styles *attached to a long beak which rises from the receptacle.* In fruit the styles split away from the beak and *curl upwards, carrying the carpels with them.*

Synopsis of the Genera.
1. Geranium. Stamens 10, all with anthers.
2. Ero'dium. Stamens with anthers, only 5.

1. GERANIUM, L. CRANESBILL.

1. **G. macula'tum,** L. (WILD C.) Stem erect, hairy, about a foot high. Leaves 5-7-parted, the wedge-shaped divisions lobed and cut. Flowers purple, an inch across. *Petals entire,* bearded on the claw, *much longer than the long-pointed sepals.*—Open woods and fields.

2. **G. Carolinia'num,** L. (CAROLINA C.) Stem usually decumbent, hairy. Sepals *awn-pointed,* as long as the *notched* rose-coloured petals.—Waste places.

3. **G. Robertia'num,** L. (HERB ROBERT.) Stems reddish, spreading, pubescent; branches weak. *Leaves 3-divided,* or *pedately 5-divided,* the divisions twice pinnatifid. Sepals awned, shorter than the reddish-purple petals. *Plant with a very strong odour.*—Shaded ravines and moist woods.

4. **G. pusil'lum,** L. (SMALL-FLOWERED C.) Stem procumbent, slender, minutely pubescent. Leaves rounded, kidney-shaped, deeply 5-7-cleft, the divisions wedge-shaped. *Sepals awnless,* about the same length as the purplish petals.—Waste places.

2. ERO'DIUM, L'Her. STORKSBILL.

E. cicuta'rium, L'Her. Stem low and spreading, hairy. *Leaves pinnate,* the leaflets sessile, pinnatifid. Peduncles several-flowered. Styles when they separate from the beak *bearded on the inside.*—Not common.

Order XXII. **OXALIDA'CEÆ.** (Wood-Sorrel F.)

Low herbs with an acid juice and alternate compound leaves, the 3 leaflets obcordate and drooping in the evening. Flowers very much the same in structure as in the preceding Order, but the fruit is a 5-celled pod, each cell opening in the middle of the back (loculicidal), and the valves persistent. Styles 5, separate. The only genus is

OX'ALIS, L. Wood-Sorrel.

1. **O. Acetosel'la**, L. (White Wood-Sorrel.) Scape 1-flowered   Petals *white, with reddish veins.*—Cold woods.

2. **O. stricta**, L. (*O. cornicula'ta*, L., *var. stricta*, Sav., in Macoun's Catalogue.) (Yellow W.) Peduncles 2–6-flowered, longer than the leaves. *Petals yellow.* Pod elongated, erect in fruit.—Copses and cultivated grounds.

Order XXIII. **BALSAMINA'CEÆ.** (Balsam Family.)

Smooth herbs, with succulent stems and simple exstipulate leaves. Flowers irregular, the sepals and petals coloured alike, *one of the coloured sepals spurred, the spur with a tail.* Stamens 5, coherent above. Pod bursting elastically, and discharging its seeds with considerable force. The only genus is

IMPA'TIENS, L. Touch-me-not. Jewel-Weed.

1. **I. fulva**, Nutt (Spotted Touch-me-not.) *Flowers orange-coloured, spotted with reddish brown.* Sac longer than broad, conical, tapering into a long recurved spur.—Cedar swamps and along streams.

2. **I. pal'lida**, Nutt. (Pale T.) Flowers *pale yellow, sparingly dotted with brown.* Sac dilated, broader than long, ending in a short spur.—Wet places.

Order XXIV. **RUTA'CEÆ.** (Rue Family.)

Shrubs, with compound *transparently-dotted* leaves, and an acrid taste. Flowers (with us) diœcious, appearing before the leaves. Stamens hypogynous, as many as the petals. Our only genus is

ZANTHOX'YLUM, Colden. PRICKLY ASH.

Z. America'num, Mill. (NORTHERN PRICKLY ASH.) TOOTH-ACHE TREE.) A prickly shrub, with yellowish-green flowers in dense umbels in the axils. Sepals obsolete or none. Petals 5. Stamens in the sterile flowers 5. Carpels 3-5, forming fleshy 1-2-seeded pods. Fruit very pungent and aromatic. Leaves pinnate, 4-5 pairs, with an odd one at the end.—Forming thickets in low grounds along streams.

ORDER XXV. ANACARDIA'CEÆ. (CASHEW FAMILY.)

Trees or shrubs, with a milky or resinous juice, and alternate leaves without dots or stipules. Sepals, petals, and stamens, each 5. Fruit a 1-seeded drupelet. The petals and stamens inserted under the edge of a disk which surrounds the base of the ovary. The only genus is

RHUS, L. SUMACH.

1. **R. typh'ina**, L. (STAGHORN SUMACH.) A small tree, 10-30 feet high, with *densely soft-hairy branches and stalks*. Flowers greenish-white, polygamous, forming a terminal thyrse. Fruit globular, *covered with crimson hairs*. Leaves pinnate, leaflets 11-31, oblong-lanceolate, serrate, pointed.—Dry hill-sides.

2. **R. glabra**, L., (SMOOTH S.) is *smooth*, and seldom exceeds 5 feet in height.

3. **R. Toxicoden'dron**, L. (POISON IVY.) Shrub about a foot high, smooth, often climbing by rootlets. Leaves 3-foliolate, leaflets rhombic-ovate, notched irregularly. Flowers polygamous, in slender axillary panicles. Plant poisonous to the touch. Var. **radi'cans**, L. has the leaves *entire*.

4. **R. venena'ta**, DC. (POISON ELDER.) A tall shrub, smooth or nearly so. Leaves odd-pinnate ; leaflets 7-13, obovate-oblong, entire. Greenish-white flowers as in No. 3.—Swamps.

5. **R. aromat'ica**, Ait., (FRAGRANT S.) is a shrub 2-3 feet high, with 3-foliolate leaves, sweet-scented when crushed, and catkin-like spikes of pale yellow flowers appearing before the leaves.—Not common.

Order XXVI. **VITA'CEÆ.** (Vine Family.)

Shrubs climbing by tendrils, with small greenish flowers in panicled clusters opposite the leaves. Stamens as many as the petals and opposite them. Calyx minute. Petals 4 or 5, hypogynous or perigynous, very deciduous. Fruit a berry, 1-4-seeded. Leaves palmately-veined, or compound.

Synopsis of the Genera.
1. **Vitis.** Leaves simple, heart-shaped, and variously lobed.
2. **Ampelop'sis.** Leaves compound-digitate, of 5 serrate leaflets.

1. VITIS, Tourn. Grape.

1. **V. æstiva'lis,** Michx. (Northern Fox-Grape.) *Leaves and branches woolly.* Berries large, dark purple or amber-coloured.—Moist thickets.

2. **V. cordifo'lia,** Michx. (Frost Grape.) *Leaves smooth* or nearly so, bright green on both sides, heart-shaped, sharply serrate. Berries small, blue or black. Var. **ripa'ria,** Michx., has broader cut-lobed leaves.—Banks of streams.

2. AMPELOP'SIS, Michx. Virginia Creeper.

A. quinquefo'lia, Michx. A common woody vine in low grounds. Leaves digitate, of 5 oblong-lanceolate leaflets. Tendrils with sucker-like disks at the end, by which they cling to walls, trunks of trees, etc. Fruit a small black berry.

Order XXVII. **RHAMNA'CEÆ.** (Buckthorn Family.)

Shrubs with simple stipulate leaves, and small regular perigynous greenish or whitish flowers. Stamens opposite the petals, and with them inserted on the margin of a fleshy disk which lines the calyx-tube. Fruit a berry-like drupe, or a pod.

Synopsis of the Genera.
1. **Rham'nus.** Petals minute, *or none.* Drupe berry-like. Calyx and disk free from the ovary.
2. **Ceano'thus.** Petals white, long-clawed, hooded. Fruit dry, dehiscent. Calyx and disk adherent to the base of the ovary.

1. RHAM'NUS, Tourn. Buckthorn.

R. alnifo'lius, L'Her. A low erect shrub, not thorny, with oval, acute, serrate leaves, and apetalous flowers. Fruit a 3-seeded berry.—Swamps.

2. CEANO'THUS, L. New Jersey Tea.

1. **C. America'nus,** L. A shrubby plant with downy branches, and ovate, 3-ribbed, serrate leaves. Flowers in white clusters at the summit of the naked flower-branches. Sepals and petals white, the latter hooded, and with slender claws. Pedicels also white.—Dry hill-sides.

2. **C. ova'lis,** Bigel., (*C. ovatus*, Desf., in Macoun's Catalogue) has the leaves narrowly oval or elliptical-lanceolate, finely serrate, and glabrous or nearly so. The flowers, also, are larger than in No. 1.—South-western Ontario.

Order XXVIII. CELASTRA'CEÆ. (Staff-tree F.)

Shrubs with simple stipulate leaves, alternate or opposite, and small regular flowers, the sepals and petals both imbricated in the bud. Stamens 4-5, alternate with the petals, and inserted on a disk which fills the bottom of the calyx. Pods orange or crimson when ripe.

Synopsis of the Genera.

1. **Euon'ymus.** Flowers perfect. Sepals 4 or 5, united at the base, and forming a *flat calyx*. Branchlets 4-sided; *leaves opposite*. *Flowers axillary*.
2. **Celas'trus.** Flowers polygamous. Petals and stamens 5. Calyx cup-shaped. *Leaves alternate*. Flowers in *terminal racemes*.

1. EUON'YMUS, Tourn. Spindle-tree.

1. **E. America'nus,** L. (Strawberry Bush.) A low, rather straggling shrub, with *short-petioled* or *sessile leaves*, the latter ovate or obovate, pointed. Flowers greenish, with the parts generally in fives. Pods *rough-warty, depressed*, crimson when ripe.—Wooded river-banks and low grounds.

2. **E. atropurpu'reus,** Jacq., (Burning Bush) occurs in the west of Ontario, and may be distinguished from No. 1 by its greater size (4-8 feet high), its *long-petioled leaves, purplish flowers*, and smooth pods.

2. CELAS'TRUS, L. Staff-tree.

C. scandens, L. (Wax-work. Climbing Bitter-Sweet.) A twining smooth shrub, with oblong-ovate, serrate, pointed leaves. Flowers small, greenish, in terminal racemes. Pods orange-coloured. These burst in autumn and display a scarlet pulpy aril, presenting a highly ornamental appearance.—Twining over bushes on river-banks and in thickets.

ORDER XXIX. **SAPINDA'CEÆ.** (SOAPBERRY FAMILY.)

Trees or shrubs, with compound or lobed leaves, and usually unsymmetrical and often irregular flowers. Sepals and petals 4-5, both imbricated in the bud. Stamens 5-10, inserted on a fleshy disk which fills the bottom of the calyx-tube. Ovary 2-3-celled, with 1 or 2 ovules in each cell.

Synopsis of the Genera.

1. **Staphyle'a.** *Flowers perfect.* Lobes of the coloured calyx, the petals, and the stamens, each 5. *Fruit a 3-celled, 3-lobed, inflated pod.* Leaves pinnately compound.
2. **Acer.** *Flowers polygamous.* Leaves simple, variously lobed, opposite. Calyx coloured, usually 5-lobed. Petals none, or as many as the sepals. Stamens 3-12. *Fruit two 1-seeded samaras* joined together, at length separating.

1. STAPHYLE'A, L. BLADDER-NUT.

S. trifo'lia, L. (AMERICAN BLADDER-NUT.) Shrub, 4-6 feet high. Leaflets 3, ovate, pointed. Flowers white, in drooping racemes, at the ends of the branchlets.—Thickets and hill-sides.

2. ACER, Tourn. MAPLE.

1. **A. Pennsylva'nicum,** L. (STRIPED MAILE.) A small tree, 10-20 feet high, with light-green bark striped with dark lines. Leaves 3-lobed at the apex, finely and sharply doubly-serrate, the lobes taper-pointed. Flowers greenish, in terminal racemes, appearing after the leaves. Samaras large, with divergent wings.—Rich woods.

2. **A. spica'tum,** Lam. (MOUNTAIN MAPLE.) A shrub or small tree, 4-8 feet high, growing in clumps in low grounds. Leaves 3-lobed, coarsely serrate, the lobes taper-pointed. Flowers greenish, appearing after the leaves, in dense *upright* racemes. Fruit with small widely-diverging wings.

3. **A. sacchari'num,** Wang. (SUGAR MAPLE.) A fine tree, with 3-5-lobed leaves, a paler green underneath, *the sinuses rounded, and the lobes sparingly sinuate-toothed.* Flowers greenish-

yellow, *drooping on slender hairy pedicels*, appearing at the same time as the leaves. Calyx fringed on the margin. Var. **nigrum**, Torr. and Gray, may be distinguished from the ordinary form by its paler and more pubescent leaves.—Rich woods.

4. **A. dasycar'pum**, Ehrhart. (WHITE or SILVER M.) Leaves deeply 5-lobed, the sinuses rather acute, silvery-white underneath, the divisions narrow, sharply-toothed. Flowers in erect clusters, greenish-yellow, *appearing much before the leaves ; petals none*. Samara very large, *woolly when young*.—River-banks and low grounds.

5. **A. ru'brum**, L. (RED M.) Leaves 3-5-lobed, the sinuses acute. *Flowers red*, appearing much before the leaves. *Petal linear-oblong.* Samara *small and smooth*, on drooping pedicels. A smaller tree than No. 4, with reddish twigs, and turning bright crimson in the autumn.—Swamps.

ORDER XXX. **POLYGALA'CEÆ**. (MILKWORT FAMILY.)

Herbs with entire exstipulate leaves, and irregular hypogynous flowers. Stamens 6 or 8, monadelphous or diadelphous, the anthers 1-celled, and opening at the top by a pore. Pod 2-celled and 2-seeded, flattened contrary to the partition. The only genus with us is

POLYGALA, Tourn. MILK WORT.

Sepals 5, the upper one and the two lower ones small and often greenish, the 2 lateral ones (called wings) larger and coloured like the petals. Petals 3, connected with each other and with the tube of filaments, the lower one keel-shaped, and usually fringed or crested at the top. Style prolonged and curved.

1. **P. verticilla'ta**, L. Flowers small, greenish-white, in slender spikes. Stems 4-8 inches high, much branched. *Stem-leaves linear, 4-5 in a whorl*, the upper ones scattered.—Dry soil.

2. **P. Sen'ega**, L. (SENECA SNAKEROOT.) Flowers greenish-white, in a solitary cylindrical close spike. Stems several, from a hard knotty rootstock, 6-12 inches high. Leaves lanceolate, with rough margins, alternate.—Dry hill-sides and thickets.

3. **P. polyg'ama**, Walt. Flowers rose-purple, showy, fringed, in a *many-flowered* raceme. Stamens 5-8 inches high, tufted and

very leafy, the leaves linear-oblong or oblanceolate. Whitish fertile flowers on underground runners.—Dry soil.

4. P. paucifo'lia, Willd. (FRINGED P.) Flowers rose-purple, very showy, fringed, *only 1-3 in number*. Stems 1-4 inches high, from long underground runners, which also bear concealed fertile flowers. Leaves ovate, crowded at the top of the stem.—Dry woods.

5. P. sanguin'ea, L. Flowers usually bright red-purple, but sometimes pale. Corolla inconspicuously crested. Flowers in *dense globular heads*, at length oblong. True petals mostly shorter than the wings, the latter broadly ovate, closely sessile. Stem leafy to the top; leaves oblong-linear.—Sandy places.

ORDER XXXI. **LEGUMINO'SÆ.** (PULSE FAMILY.)

Herbs, shrubs, or trees, mostly with compound alternate stipulate leaves, and papilionaceous corollas. (For description of a typical flower, see Part I., cap. v.) Stamens usually 10, monadelphous, diadelphous, or distinct. Fruit a legume.

Synopsis of the Genera.
* *Flowers papilionaceous.*

1. **Lupi'nus.** Leaves palmately-compound, leaflets 7-9. Flowers in terminal racemes. Stamens monadelphous.
2. **Trifo'lium.** Leaves of 3 leaflets. *Flowers in heads.* Stamens diadelphous.
3. **Medica'go.** Leaves pinnate, of 3 leaflets. Flowers in axillary spikes. *Pod curved or coiled.* Stamens diadelphous.
4. **Melilo'tus.** Leaves pinnate, of 3 leaflets, the leaflets toothed. Flowers in slender axillary racemes. Pod wrinkled, 1-2-seeded. Stamens diadelphous.
5. **Robin'ia.** *Trees.* Leaves odd-pinnate, often with spines for stipules, and the leaflets with small stipules. Flowers in hanging axillary racemes. Pod margined on one edge. Stamens diadelphous.
6. **Astrag'alus.** Leaves odd-pinnate, leaflets numerous. Flowers in dense axillary spikes. Corolla long and narrow. Pod turgid, *one or both sutures* (see Part I., section 217) *projecting into the cell, thus partially or wholly dividing the cavity.* Stamens diadelphous.
7. **Desmo'dium.** Leaves pinnate, of 3 leaflets. *Calyx 2-lipped.* Flowers purple or purplish, in axillary or terminal racemes. Pod flat, *the lower margin deeply lobed, thus making the pod jointed*, roughened with hooked hairs, causing the pods to adhere to the clothing, etc. Stamens diadelphous.
8. **Lespede'za.** Leaves pinnate, of 3 leaflets. *Calyx 5-cleft.* Pod flat, oval or roundish, *occasionally 2-jointed, but only 1-seeded.* Flowers sometimes polygamous. Stamens diadelphous.

9. **Vicia.** Leaves abruptly pinnate, *the leafstalk prolonged into a tendril.* Flowers axillary. *Style filiform, hairy at the apex.* Pod 2-several-seeded. Stamens diadelphous.

10. **Lath'yrus.** Leaves as in Vicia. *Style flattish,* flattened above, an hairy down the side opposite the free stamen. Stamens diadelphous.

11. **A'pios.** *A twining herb. Leaves pinnate, 5-7 leaflets.* Keel of the slender and *coiled inward.* Flowers in dense racemes. Stamens d phous.

12. **Amphicarpæ'a.** *A low and slender twiner,* the stem clothed brownish hairs. *Leaves pinnate, of 3 leaflets.* Flowers polygan those of the upper racemes perfect, those near the base fertile, wit corolla inconspicuous or none. Stamens diadelphous.

13. **Baptis'ia.** Leaves palmate, of 3 leaflets. *Stamens all separate.* The *keel-petals nearly separate.* Racemes terminating the bushy branches

** *Flowers not papilionaceous; polygamous. Trees.*

14. **Gledit'schia.** Thorny trees, with abruptly once- or twice-pinnate leaves. Flowers greenish, inconspicuous, in small spikes.

1. LUPI'NUS, Tourn. LUPINE.

L. peren'nis, L. (WILD LUPINE.) Stem erect, somewha hairy. Leaflets 7-9, oblanceolate. Calyx deeply 2-lipped. Pods hairy.—Sandy soil.

2. TRIFO'LIUM, L. CLOVER. TREFOIL.

1. **T. arvense, L.** (RABBIT-FOOT or STONE CLOVER.) Stem erect, 4-12 inches high, branching. Heads of whitish flowers oblong, *very silky and soft.* Calyx-teeth fringed with long silky hairs.—Dry fields.

2. **T. pratense, L.** (RED C.) Stems and leaves somewhat hairy, the latter marked with a pale spot on the upper side. Flowers purplish, in dense heads.—Pastures.

3. **T. repens, L.** (WHITE C.) Smooth, creeping. Heads of white flowers rather loose.—Fields everywhere.

4. **T. reflexum, L.** (BUFFALO C.) Only in south-western Ontario, in the neighborhood of the Detroit river. Heads large, on naked peduncles; standard rose-red, wings and keel whitish. Flowers reflexed when old.

5. **T. agrarium, L.** (YELLOW or HOP-C.) Flowers *yellow,* reflexed when old. Leaflets obovate-oblong, *all 3 from the same point.* Stem 6-12 inches high.—Sandy fields.

6. **T. procumbens,** L. (Low Hop-C.) Flowers yellow, reflexed when old. Leaflets *wedge-obovate*, the lateral ones at a short distance from the terminal one. Stem smaller than in No. 5, trailing.—Sandy fields.

3. MEDICA'GO, L. MEDICK.

1. **M. lupuli'na,** L. (BLACK MEDICK.) Stem procumbent, y. Leaflets obovate, toothed at the apex. Flowers yellow. kidney-shaped.—Waste places.

2. **M. sati'va,** L., (LUCERNE) has *purple flowers* in a long e, and *spirally-twisted pods.*—Cultivated fields.

4. MELILO'TUS, Tourn. SWEET CLOVER.

1. **M. officina'lis,** Willd. (YELLOW MELILOT.) Stem erect, 4 feet high. Leaflets obovate-oblong. Flowers yellow. Pod drooping, 1-2-seeded.—Waste places.

2. **M. alba,** Lam., (WHITE M.) is much like No. 1, but has te flowers.—Escaped from gardens.

5. ROBIN'IA, L. LOCUST-TREE.

1. **R. Pseudaca'cia,** L. (COMMON LOCUST.) *Racemes slender, loose.* Flowers white, fragrant. A large tree.

2. **R. visco'sa,** Vent. (CLAMMY L.) *Racemes crowded.* Flowers white, with a reddish tinge. *Branchlets and leafstalks clammy.* Smaller than No. 1.

6. ASTRAG'ALUS, L. MILK-VETCH.

1. **A. Canaden'sis,** L. (CANADIAN MILK-VETCH.) Stem erect, 1-4 feet high, somewhat pubescent. Leaflets *10 or more pairs*, with an odd one at the end. *Flowers greenish-yellow, very numerous.*—River-banks.

2. **A. Coop'eri,** Gray, has *fewer leaflets*, and *white flowers* in a short spike.—Not common.

7. DESMO'DIUM, DC. TICK-TREFOIL.

* *Pod raised on a stalk much surpassing the calyx, the latter slightly toothed. Stipules bristle-form.*

1. **D. nudiflo'rum,** DC. Stem smooth, 4-8 inches high. Leaves crowded at the summit of sterile stems. Flowers in a terminal raceme or panicle, *on a scape which rises from the root.* Leaflets broadly ovate.

2. **D. acumina'tum.**, DC. Stem pubescent. Leaves all crowded at the summit of the stem, *from which the raceme or panicle arises* Leaflets conspicuously pointed.— Rich woods.

3. **D. pauciflo'rum**, DC. Leaves *scattered* along the low ascending stems; leaflets rhombic-ovate, rather blunt. Racemes few flowered, terminal.—Rich woods, western Ontario.

** *Pod raised on a stalk hardly surpassing the calyx, the latter deeply cleft Stipules ovate, taper-pointed.*

4. **D. rotundifo'lium**, DC. Stem *prostrate, soft-hairy*. Leafl. orbicular. Flowers purple. Pods indented on both edges. Dry sandy woods, western Ontario.

*** *Pod hardly, if at all, stalked.*

5. **D. cuspida'tum**, Torr. and Gray. Stem tall, erect, very smooth. Leaflets ovate-lanceolate, taper-pointed, very large, green on both sides. Flowers and bracts large. Pod 4-6-jointed —Thickets.

6. **D. panicula'tum**, DC. Stem slender, nearly smooth eaflets oblong-lanceolate, tapering to a blunt point. Flowe, medium-sized. Pod 3-5-jointed, the joints triangular. Race. panicled.—Rich woods.

7. **D. Dille'nii**, Darlingt. Distinguished from the last by the *pubescent stem* and finely pubescent leaflets, the latter *oblong* or *oblong-ovate*.—Dry and open thickets.

8. **D. Canadense**, DC. Stem erect, *hairy*, tall, furrowed. Leaflets oblong-lanceolate (1½-3 inches long), *with many straightish veins*. Flowers large, about ½ inch long, in dense racemes. Joints of the pod roundish. —Dry woods.

9. **D. cilia're**, DC. Stem ascending, slender, hairy. Leaflets round-ovate (½-1 inch long). Flowers small, in *loose racemes*.— Dry thickets, south-western Ontario.

8. LESPEDE'ZA. Bush-Clover.

Flowers of two sorts; the larger perfect, the smaller pistillate and usually apetalous, mingled with the others.

1. **L. viola'cea**, Pers. (*L. reticulata*, Pers., in Macoun's Catalogue.) Stems upright, branched. Leaflets varying from oblong to linear, downy underneath. *Flowers violet-purple.*—Dry borders of woods, western Ontario.

** *All the flowers perfect, in close spikes or heads.*

2. **L. hirta,** L. Stem erect, wand-like, tall, pubescent. Leaflets roundish or oval, pubescent. Spikes dense, on *peduncles larger than the leaves*. Corolla *yellowish-white*, with a purple spot on the standard.

3. **L. capita'ta,** Michx. *Peduncles and petioles short.* Leaflets varying from oblong to linear, silky underneath. Flowers in dense heads; *corolla as in No. 1.* Calyx much longer than the pod.—Both species are found in dry soil.

9. VICIA, Tourn. VETCH.

V. sati'va, L. (COMMON VETCH or TARE.) Stem simple, somewhat pubescent. Leaflets 10-14, varying from obovate-oblong to linear. *Flowers purple, large, one or two together, sessile in the axils, or nearly so.*—Cultivated fields and waste grounds.

2. **V. Cracca,** L. (TUFTED V.) Downy-pubescent. *Leaflets 0-24*, oblong-lanceolate, strongly mucronate. *Peduncles long, bearing a dense one-sided raceme of blue flowers*, bent downward in the spike, and turning purple before withering.—Borders of thickets, and pastures. Chiefly eastward.

3. **V. Carolinia'na,** Walt. Smooth. Leaflets 8-12, oblong. Peduncles bearing a *rather loose raceme of whitish flowers, the keel tipped with blue.*—Low grounds and river-banks.

4. **V. America'na,** Muhl. Smooth. *Leaflets 10-14*, oval or ovate-oblong, very veiny. *Peduncles 4-8-flowered, flowers purple.*—Moist places.

5. **V. hirsu'ta,** Koch. Stem weak. *Leaflets 12-16, linear.* Peduncles 3-6-flowered. *Pods hairy, 2-seeded.*—Chiefly eastward.

10. LATH'YRUS, L. EVERLASTING PEA.

1. **L. marit'imus,** Bigel. (BEACH PEA.) Stem stout, about a foot high. Leaflets 8-16, oval or obovate. *Stipules broadly halberd-shaped, about as large as the leaflets.* Flowers large, purple.—Sea-coast, and shores of the Great Lakes.

2. **L. veno'sus,** Muhl. (VEINY E.) Stem 2-3 feet high. Leaflets 10-14. *Stipules very small, slender, half arrow-shaped.* Flowers numerous.—Shady banks, chiefly westward and southward.

3. **L. ochroleu'cus**, Hook. (Pale E.) Stem slender. 1
6-8, smooth and glaucous. *Stipules half heart-shaped,*
Corolla yellowish-white.—Chiefly northward.

4. **L. palus'tris**, L. (Marsh E.) Stem slender,
margined. Leaflets 4-8, lanceolate, linear, or narrowly ob
sharply mucronate. Stipules small, half arrow-shaped. C
blue-purple.—Moist places Var. **myrtifolius** has oblong-l
late leaflets, and pale-purple flowers. Upper stipules
larger than the lower ones.

A'PIOS, Boerhaave. Ground-Nut. Wild Be·

A. tubero'sa, Mœnch. Flowers brown-purple.—A
twining plant in low grounds.

12. **AMPHICARPÆ'A**, Ell. Hog Pea-Nut.

A. monoi'ca, Nutt. Flowers white or purplish.—Moi
thickets and river-banks.

13. **BAPTIS'IA**, Vent. False Indigo.

B. tincto'ria, R. Br. (Wild Indigo.) Smooth and slender,
2-3 feet high, branching. Leaves nearly sessile. Leaflets wedg
obovate, turning black on drying. Flowers yellow.—Dry soil.
Lake Erie coast.

14. **GLEDIT'SCHIA**, L. Honey Locust.

1. **G. triacan'thos**, L. Thorns stout, often triple or compound. Pods linear, often more than a foot long, with pulp
between the flat seeds.—Common in cultivation, and established
on Point Pelee.

Order XXXII. **ROSA'CEÆ**. (Rose Family.)

Herbs, shrubs, or trees, with alternate stipulate leaves, and
regular flowers. The petals (mostly 5) and stamens (mostly more
than 10) inserted on the edge of a disk which lines the calyx-
tube. (See Part I., sections 48 to 57, for typical flowers.)

Synopsis of the Genera.

Suborder **AMYGDALEÆ**.

1 **Pru'nus**. Calyx 5-cleft, free from the ovary, deciduous. Fruit a drupe.

Suborder ROSACEÆ.

___'a. Carpels mostly 5, forming follicles in fruit. Calyx 5-cleft, short. tals obovate, similar.

.le'nia. Carpels and fruit as in Spiræa. Calyx elongated, 5-toothed. 'etals slender, dissimilar.

grimo'nia. Carpels 2, forming achenes enclosed in the hardened calyx ibe. Calyx armed with hooked bristles. Flowers yellow, in slender ikes.

___ ___ 'ia. Carpels numerous, one-ovuled, becoming dry achenes, the persistent styles becoming tails, plumose or naked, and straight or jointed. lyx-lobes with 5 alternating bractlets.

stei'nia. Carpels 2-6, forming achenes. Leaves radical, of 3 wedge-leaflets. Bractlets of the calyx minute and deciduous. Flowers ___ow, on bracted scapes.

___ entil'la. Carpels numerous, forming achenes heaped on a dry receptacle, the styles not forming tails. Lobes of the calyx with 5 alternating bracts.

3. Fraga'ria. Flowers as in Potentilla, but receptacle becoming fleshy or pulpy and scarlet in fruit. (See Part I., section 235.) Leaves all radical, of 3 leaflets. Low plants producing runners.

. Dalibar'da. Carpels 5-10, each 2-ovuled, forming nearly dry drupelets. Calyx 5-6-parted, 3 of the divisions larger than the others, and toothed. Calyx without bracts, persistent, enclosing the fruit. Leaves radical, round heart-shaped. Flowers white, on scapes.

10. Rubus. Carpels numerous, 2-ovuled, forming drupelets heaped on the receptacle. (See Part I., section 234.) Fruit edible. Calyx without bracts.

11. Rosa. Carpels numerous, 1-ovuled, forming achenes enclosed in the fleshy calyx-tube. (See Part I., section 49.)

Suborder POMEÆ.

12. Cratæ'gus. Calyx-tube urn-shaped, becoming thick and fleshy in fruit, enclosing and combined with the 2-5 carpels. Fruit a pome, but drupe-like, containing 2-5 bony nutlets. *Thorny shrubs.* Flowers generally white.

13. Pyrus. Fruit a pome or berry-like, the 2-5 carpels or cells of a papery or cartilaginous texture (see Part I., sections 52 and 232), each 2-seeded. Shrubs or trees.

14. Amelan'chier. Pome berry-like, *10-celled, i.e.*, with twice as many cells as styles. Petals narrow. Otherwise as in Pyrus. Shrubs or small trees, *not thorny.*

1. PRUNUS. Tourn. PLUM. CHERRY.

1. P. Americ'a'na, Marshall. (WILD PLUM.) A thorny tree 8-20 feet high, with orange or red drupes half an inch or more in diameter; and ovate, conspicuously pointed, serrate, veiny leaves. Flowers white, appearing before the leaves, in umbel-like lateral clusters.—Woods and river-banks.

2. **P. pu'mila**, L. (DWARF CHERRY.) A small trailing shrub, 6-18 inches high. Leaves obovate-lanceolate, tapering to the base, toothed near the apex, pale beneath. Flowers in umbels of 2-4, appearing with the leaves. Fruit ovoid, dark red, as large as a good-sized pea.—Sandy or gravelly soil, along the Great Lakes.

3. **P. Pennsylva'nica**, L. (WILD RED CHERRY.) A tree 20-30 feet high, or shrubby. Leaves oblong-lanceolate, sharply serrate, green both sides. Flowers (appearing with the leaves) in large clusters, the pedicels elongated. Fruit globular, as large as a red currant, very sour.—Rocky thickets, and in old windfalls.

4. **P. Virginia'na**, L. (CHOKE-CHERRY.) A good-sized shrub, 3-10 feet high. Leaves oval, oblong, or obovate, finely and sharply serrate, abruptly pointed. Flowers in short erect racemes, appearing after the leaves. Fruit red, becoming darker, very astringent.—Woods and thickets.

5. **P. sero'tina**, Ehrhart. (WILD BLACK CHERRY.) A large tree, with reddish-brown branches. Leaves smooth, varying from oval to ov te lanceolate, taper-pointed, serrate, with short and blunt incurved teeth, shining above. Flowers in long racemes. Fruit purplish-black, edible.—Woods and thickets.

2. SPIRÆ'A, L. MEADOW-SWEET.

1. **S. opulifo'lia**, L. (*Neillia opulifolia*, Benth. and Hook., in Macoun's Catalogue.) (NINE-BARK.) Shrub 3-7 feet high, the old bark separating in thin layers. Leaves broadly ovate or cordate, 3-lobed, doubly crenate, smooth. Flowers white, in umbel-like corymbs terminating the branches. Follicles 2-5, inflated, purplish.--River-banks.

2. **S. salicifo'lia**, L. (COMMON MEADOW-SWEET.) Shrub 2-3 feet high, nearly smooth. Leaves wedge-lanceolate, doubly serrate. Flowers white or rose-coloured, in a dense, terminal panicle.—Low grounds along streams.

3. **S. tomento'sa,** L., (DOWNY M.) with deep rose-coloured flowers, and the stems and under surface of the leaves densely woolly, occurs eastward towards the sea-coast, and in the northern counties of Ontario.

3. GILLE'NIA, Mœnch. INDIAN-PHYSIC.

G. trifolia'ta, Mœnch. (BOWMAN'S ROOT.) Herb with 3-foliolate leaves; the leaflets ovate-oblong, pointed, rather coarsely serrate; stipules small, awl-shaped, entire. Flowers white or rose-coloured, in loose few-flowered corymbs.—Rich woods, chiefly south-westward.

4. AGRIMO'NIA, Tourn. AGRIMONY.

A. Eupato'ria, L. (COMMON AGRIMONY.) Stem herbaceous, hairy, 2-3 feet high. Leaves interruptedly pinnate, larger leaflets 5-7, oblong-obovate, coarsely serrate. Petals yellow, twice as long as the calyx.—Borders of woods.

5. GEUM, L. AVENS.

1. **G. album,** Gmelin. (WHITE AVENS.) Stem 2 feet high, slender, branching, smoothish or downy. Root-leaves pinnate, the cauline ones 3-divided, lobed, or only toothed. *Petals white,* as long as the calyx. Achenes bristly, tipped with the hooked lower joint of the style, the upper joint falling away. *Receptacle of the fruit bristly.*--Low rich woods and thickets.

2. **G. Virginia'num,** L. Stem stout, bristly-hairy. Leaves nearly as in No. 1. *Petals white, shorter than the calyx. Receptacle of the fruit nearly smooth.*—Meadows and thickets; not common.

3. **G. strictum,** Ait. (YELLOW A.) Stem 2-3 feet high, rather hairy. Root-leaves interruptedly pinnate; stem-leaves 3-5 foliolate, leaflets obovate or ovate. *Petals yellow,* longer than the calyx. Receptacle of the fruit downy. Achenes tipped with the hooked style.—Dry thickets.

4. **G. riva'le,** L. (WATER or PURPLE AVENS.) *Petals purplish-yellow; calyx brown-purple.* Flowers nodding, but the fruiting heads upright. *The upper joint of the style feathery, persistent.* Stem simple, 2 feet high. Root-leaves lyrate; stem-leaves few, 3-foliolate, lobed.—Bogs and wet places.

5. **G. triflo'rum,** Pursh. Stem about a foot high, soft-hairy. Flowers 3 or more, on long peduncles, purple. Styles not jointed, feathery, *at least 2 inches long in the fruit.*—Dry hills and thickets; not common.

6. **WALDSTEI'NIA,** Willd. BARREN STRAWBERRY.

W. fragarioi'des, Tratt. A low plant, 4-6 inches high. Leaflets 3, broadly wedge-form, crenately toothed. Scapes several-flowered. Petals yellow, longer than the calyx.—Dry woods and hill-sides.

7. **POTENTIL'LA.** L. CINQUE-FOIL.

1. **P. Norve'gica,** L. (NORWAY CINQUE-FOIL.) Stem *erect, hairy,* branching above. *Leaves palmate, of 3 leaflets;* leaflets obovate-oblong, coarsely serrate. Flowers in cymose clusters. Petals pale yellow, small, *not longer than the sepals.*—Fields and low grounds.

2. **P. paradox'a,** Nutt., a plant of spreading or decumbent habit, with *pinnate leaves of 5-9 leaflets,* solitary flowers, *small petals,* and achenes with an appendage at the base, occurs along the south-western shore of Lake Ontario.

3. **P. Canaden'sis,** L. (CANADA C.) Stem prostrate or ascending, silky-hairy. *Leaves palmate, of 5 leaflets,* the latter serrate towards the apex. Flowers solitary. Petals yellow, *longer than the sepals.*—Dry soil.

4. **P. argen'tea,** L. (SILVERY C.) Stem ascending, branched at the summit, *white-woolly.* Leaves palmate, of 5 leaflets, the latter deeply serrate towards the apex, *with revolute margins, and woolly beneath.* Petals yellow, longer than the sepals.—Dry fields and roadsides.

5. **P. argu'ta,** Pursh. Stem stout, 1-2 feet high, brownish-hairy. Leaves pinnate, of 3-9 oval serrate leaflets, downy underneath. Flowers in dense cymose clusters. Petals yellowish or cream-coloured, deciduous. Plant clammy above.—Dry thickets.

6. **P. Anseri'na,** L. (SILVER-WEED.) A low plant, *creeping with slender runners.* Leaves all radical, interruptedly pinnate; leaflets 9-19, serrate, green above, *silvery-silky beneath. Flowers solitary, on long scape-like peduncles,* bright yellow.—River and lake margins.

7. **P. frutico'sa,** L. (SHRUBBY C.) Stem erect, *shrubby,* 1-3 feet high, much branched. Leaves pinnate, of 5-7 leaflets, closely crowded, *entire,* silky, especially beneath. Flowers numerous, large, yellow, terminating the branches.—Bogs.

8. **P. tridenta'ta,** Ait., (THREE-TOOTHED C.) is common eastward towards the sea-coast. Stem 4-6 inches high. Leaves rigid, palmate, of 3 wedge-shaped leaflets, *3-toothed at the apex. Petals white.*

9. **P. palustris,** Scop. (MARSH FIVE-FINGER.) Stem ascending. Leaves pinnate, of 5-7 lanceolate, crowded, deeply serrate leaflets, whitish beneath. *Calyx an inch broad, dark purple inside. Petals purple.*—Bogs.

8. FRAGA'RIA, Tourn. STRAWBERRY.

1. **F. Virginia'na,** Ehrhart. *Achenes deeply imbedded in pits* on the surface of the fleshy receptacle; calyx erect after flowering. Leaflets firm.

2. **F. ves'ca,** L. *Achenes not sunk in pits,* but merely on the surface of the receptacle; calyx spreading. Leaflets thin.

9. DALIBAR'DA, L. DALIBARDA.

D. repens, L. (*Rubus Dalibarda,* L., in Macoun's Catalogue.) Stems tufted, downy. Whole plant with something of the aspect of a violet.—Low woods.

10. RUBUS, Tourn. BRAMBLE.

1. **R. odora'tus,** L. (PURPLE FLOWERING-RASPBERRY.) Shrubby, 3-5 feet high. Branches, peduncles, and calyx *clammy with glandular hairs. Flowers large and handsome, rose-purple.* Leaves large, broadly ovate, 3-5 lobed, the lobes acute, minutely toothed. *Fruit flat.*

2. **R. triflo'rus,** Richardson. (DWARF RASPBERRY.) Stems ascending or trailing, a foot high, not prickly. Leaflets 3-5, nearly smooth, rhombic-ovate, acute at both ends, doubly serrate. Peduncle usually 3-flowered. Petals white; sepals reflexed. Fruit red.—Cedar swamps.

3. **R. strigo'sus,** Michx. (WILD RED RASPBERRY.) *Stems upright, beset with stiff straight bristles.* Leaflets 3-5, oblong-ovate, pointed, cut-serrate, whitish beneath. *Fruit light red.*—Hillsides and thickets.

4. **R. occidenta'lis,** L. (BLACK RASPBERRY.) *Stem glaucus, recurved, armed with hooked prickles.* Leaflets 3, ovate, pointed, coarsely serrate, white-downy beneath. Fruit purplish-black.—Borders of fields, especially where the ground has been burned over.

5. **R. villo'sus,** Ait. (HIGH BLACKBERRY.) Stem shrubby, furrowed, erect or reclining, armed with hooked prickles. Leaflets 3–5, unequally serrate, the terminal one conspicuously stalked. Lower surface of the leaflets *hairy and glandular.* Flowers racemed, numerous, large and white. *Fruit oblong, black.* Var. **frondosus** is smoother and less glandular. Var. **humifusus** is trailing and smaller, and the flowers are less numerous.—Borders of thickets.

6. **R. Canaden'sis,** L. (LOW BLACKBERRY. DEWBERRY.) Stem shrubby, *extensively trailing, slightly prickly.* Leaflets chiefly 3, oval or ovate-lanceolate, *nearly smooth,* sharply serrate. Flowers in racemes.—Thickets and rocky hills.

7. **R. his'pidus,** L., (RUNNING SWAMP BLACKBERRY) occurs occasionally in low meadows. Stem prostrate, with small reflexed prickles, sending up at intervals the short flowering shoots. Leaflets mostly 3, smooth and shining. Fruit of few grains, red or purple.

11. ROSA, Tourn. ROSE.

* *Styles separate; included within the calyx-tube.*

1. **R. Caroli'na,** L. (SWAMP ROSE.) Stem 4–8 feet high, erect, armed with *stout hooked prickles,* but no bristles. Leaflets 5–9, finely serrate. *Flowers in corymbs, numerous. Calyx and globular calyx-tube beset with glandular bristles.*—Wet places.

2. **R. lu'cida,** Ehrhart. (DWARF WILD ROSE.) Stem 1–2 feet high, armed with slender *almost straight* prickles, and bristles. Leaflets 5–9, finely serrate. *Peduncles 1–3-flowered. Calyx-teeth bristly, but the tube in fruit nearly smooth.*—Dry soil, or borders of swamps.

3. **R. blanda,** Ait. (EARLY WILD ROSE.) Stem 1–3 feet high, *Prickles few and scattered, straight.* Leaflets 5–7. *Peduncles 1–3-flowered. Calyx and fruit smooth,* the lobes of the calyx erect and connivent in fruit.—Dry woods and fields.

4. **R. rubigino'sa,** L. (SWEET-BRIER.) Stem tall. Prickles numerous, the larger hooked, the smaller awl-shaped. Leaflets 5-7, doubly serrate, *glandular beneath*. *Flowers mostly solitary.* Fruit *pear-shaped* or obovate.—Roadsides and fields.

* * *Styles cohering in a protruding column, as long as the stamens.*

5. **R. setig'era,** Michx. *Stem climbing.* Prickles nearly straight. Leaflets 3-5, ovate. Petals deep rose-coloured, changing to white.—Borders of thickets and along fences; south-western Ontario.

12. CRAT.E'GUS, L. HAWTHORN.

1. **C. coccin'ea,** L. (SCARLET-FRUITED THORN.) A low tree, *glabrous*. Leaves rather *thin*, roundish-ovate, serrate, *on slender petioles*. Fruit bright red, ovoid, hardly edible.—Thickets.

2. **C. tomento'sa,** L. (BLACK or PEAR THORN.) A tall shrub or low tree, *downy*, at least when young. Leaves *thickish*, oval or broadly ovate, finely serrate, on *margined petioles*, furrowed along the veins. Fruit globular or pear-shaped, larger than in No. 1, edible.—Thickets.

3. **C. Crus-galli,** L. (COCKSPUR THORN.) A shrub or low tree, *glabrous*. Leaves thick, *shining above, wedge-obovate*, finely serrate. *Petioles very short.* Fruit globular, bright red. Thorns very long.—Thickets.

13. PYRUS, L. PEAR. APPLE.

1. **P. corona'ria,** L. (AMERICAN CRAB-APPLE.) A small tree, with ovate serrate *simple leaves*. Flowers in umbel-like cymes. Styles woolly and cohering at the base. *Fruit a greenish apple.* —Chiefly west of Toronto.

2. **P. arbutifo'lia,** L. (CHOKE-BERRY.) A shrub, with obovate finely serrate *simple leaves*. Flowers in compound cymes. *Fruit berry-like*, nearly globular, *dark red or black.*—Swamps.

3. **P. America'na,** DC. (AMERICAN MOUNTAIN ASH.) A small tree, with *odd-pinnate leaves of 13-15 leaflets*, the latter lanceolate, taper-pointed, sharply serrate, bright green. *Fruit scarlet, berry-like.* Flowers in flat cymes.—Swamps and cool woods, northward.

14. AMELAN'CHIER, Medic. JUNE-BERRY.

A. Canaden'sis, Torr. and Gray. (SHADBUSH.) A shrub or small tree, with a purplish, berry-like, edible fruit. The variety

Botryapium has ovate-oblong leaves, very sharply serrate, and white flowers in long drooping racemes, the petals 4 times as long as the calyx. The variety **rotundifolia** has broader leaves and shorter petals, and 6-10-flowered racemes.

ORDER XXXIII. SAXIFRAGA'CEÆ. (SAXIFRAGE F.)

Herbs or shrubs, distinguished from Rosaceæ chiefly in having opposite as well as alternate leaves, and usually no stipules; stamens only as many or twice as many as the (usually 5) petals; and the carpels fewer than the petals (mostly 2), and usually more or less united with each other. Stamens and petals generally inserted on the calyx.

Synopsis of the Genera.

1. **Ri'bes.** Shrubs, sometimes prickly, with alternate and palmately-veined and lobed leaves, which are plaited in the bud. Calyx 5-lobed, the tube adherent to the ovary (superior). Petals 5, small, inserted on the calyx. Stamens 5. Styles 2. *Fruit a many-seeded berry.*
2. **Parnas'sia.** Smooth herbs, with entire and chiefly radical leaves, and solitary flowers terminating the long scapes. Petals 5, large, *veiny, each with a cluster of sterile filaments at the base.* Proper stamens 5. Stigmas 4. Pod 4-valved. Calyx free from the ovary.
3. **Saxif'raga.** Herbs with *clustered root-leaves.* Flowers in close cymes. Calyx-lobes hardly adherent to the ovary. Petals 5. Stamens 10. *Fruit a pair of follicles,* slightly united at the base.
4. **Mitel'la.** Low and slender herbs, with round heart-shaped radical leaves, *those on the scape (if any) opposite.* Flowers in terminal racemes. Calyx 5-lobed, adherent to the base of the ovary. Petals 5, slender, *pinnatifid.* Stamens 10, *short.* Styles 2. Pod 2-beaked, but 1-celled.
5. **Tiarel'la.** Slender herbs, with radical heart-shaped leaves, and *leafless* scapes, bearing a simple raceme of flowers. Calyx bell-shaped, 5-parted. Petals 5, *entire.* Stamens 10, *long and slender.* Pod 2-valved, *the valves unequal.*
6. **Chrysople'nium.** Small and smooth herbs, with mostly opposite roundish leaves. Calyx-tube adherent to the ovary. *Petals none.* Stamens twice as many as the calyx-lobes (8-10), inserted on a conspicuous disk. Pod 2-lobed.

1. RI'BES, L. CURRANT. GOOSEBERRY.

1. **R. Cynos'bati,** L. (WILD GOOSEBERRY.) Stem with small thorns at the bases of the leaves, the latter downy, on slender petioles, roundish heart-shaped, 3-5-lobed. *Peduncles slender,* 2-3-flowered. *Berry covered with long prickles.*—Open woods and clearings.

2. **R. hirtel'lum,** Michx. (SMALL WILD GOOSEBERRY.) Stems with very short thorns or none. *Peduncles very short,* 1-2-flowered. *Berry small, smooth.*—Low grounds.

3. **R. lacus'tre,** Poir. (SWAMP GOOSEBERRY.) Shrubby. Young stems prickly, and thorny at the bases of the leaves. Leaves cordate, *deeply 3-5-lobed, the lobes deeply cut. Racemes 4-9-flowered, slender, nodding.* Fruit bristly.—Swamps and wet woods.

4. **R. flor'idum,** L. (WILD BLACK CURRANT.) *Stems and fruit without prickles or thorns. Leaves resinous-dotted,* sharply 3-5-lobed, doubly serrate. Racemes many-flowered, drooping. Calyx bell-shaped. *Fruit black,* smooth.—Woods.

5. **R. rubrum,** L. (WILD RED CURRANT.) A low shrub with straggling stems. Leaves obtusely 3-5-lobed. *Racemes from lateral buds separate from the leaf-buds,* drooping. Calyx flat. *Fruit red,* smooth.—Bogs and wet woods.

2. **PARNAS'SIA,** Tourn. GRASS OF PARNASSUS.

P. Carolinia'na, Michx. Petals sessile, very veiny. Sterile filaments 3 in each set. Leaves ovate or rounded, *usually only one low down on the stalk.* Flower an inch across.—Beaver meadows and wet banks.

3. **SAXIF'RAGA,** L. SAXIFRAGE.

S. Virginien'sis, Michx. (EARLY SAXIFRAGE.) Stem 4-9 inches high. Scape clammy. Leaves obovate, crenately toothed. Petals white, oblong, twice as long as the sepals.—Damp rocks along streams.

4. **MITEL'LA,** Tourn. MITRE-WORT. BISHOP'S-CAP.

1. **M. diphyl'la,** L. (TWO-LEAVED MITRE-WORT.) Stem hairy. Leaves cordate, 3-5-lobed, *those on the scape 2, opposite,* nearly sessile. Flowers white.—Rich woods.

2. **M. nuda,** L. (NAKED-STALKED M.) Stem small and delicate. Leaves kidney-shaped, *doubly crenate. Scape leafless,* few-flowered. *Flowers greenish.*—Deep woods, on moss-covered logs, etc.

5. **TIAREL'LA,** L. FALSE MITRE-WORT.

T. cordifo'lia, L. Scape leafless, 5-12 inches high. Leaves heart-shaped, sharply toothed, sparsely hairy above, downy beneath. Petals white, oblong.—Rich woods.

6. **CHRYSOPLE'NIUM**, Tourn. Golden Saxifrage.

C. America'num, Schwein. A low and delicate smooth herb, with spreading and forking stems. Flowers greenish-yellow, inconspicuous, nearly sessile in the forks. Shady wet places.

Order XXXIV. CRASSULA'CEÆ. (Orpine Family.)

Succulent herbs (except in one genus), chiefly differing from Saxifragaceæ in having *symmetrical flowers*, the sepals, petals, and carpels being the same in number, and the stamens either as many or twice as many.

1. PEN'THORUM, Gronov. Ditch Stone-crop.

P. sedoi'des, Gronov. Not succulent. Sepals 5. Petals 5, if any; sometimes wanting. Stamens 10. *Pod 5-angled, 5-horned, and 5-celled.* Leaves scattered, lanceolate, acute at both ends. A homely weed, with greenish-yellow flowers in a loose cyme.—Wet places. (Parts of the flowers occasionally in sixes or sevens.)

2. SEDUM, Tourn. Stone-crop. Orpine.

S. acre, L. (Mossy Stone-crop.) Leaves very thick and succulent, crowded, very small. Petals yellow. A spreading moss-like plant, which has escaped from cultivation in many places.—Roadsides.

Order XXXV. HAMAMELA'CEÆ. (Witch-Hazel F.)

Tall shrubs, with alternate simple leaves, and deciduous stipules. Flowers in clusters or heads, often monœcious. Calyx 4-parted, adherent to the base of the ovary, the latter of 2 united carpels. Fruit a 2-beaked, 2-celled, woody pod, opening at the top. Petals 4, strap-shaped, inserted on the calyx. Stamens 8, 4 of them anther-bearing, the remainder reduced to scales. The only genus with us is

HAMAME'LIS, L. Witch-hazel.

H. Virgin'ica, L. Leaves obovate or oval, crenate or wavy-toothed, pubescent. Flowers yellow, appearing late in autumn.—Damp woods, chiefly west of Toronto.

Order XXXVI. **HALORA'GEÆ.** (Water-Milfoil F.)

Aquatic or marsh plants, with small inconspicuous flowers, sessile in the axils of the leaves or bracts. Calyx-tube adherent to the ovary, the latter in one genus 4-lobed and 4-celled; in the other of a single carpel. Limb of the calyx minute or none. Petals 4, if any. Stamens 1-8. Fruit indehiscent, a single seed in each cell.

Synopsis of the Genera.

1. Myriophyl'lum. Flowers monœcious or polygamous, with the parts in fours. Stamens 4 or 8. Immersed leaves pinnately dissected into capillary divisions.
2. Hippu'ris. Flowers perfect. Stamen, style, and carpel *only one*. Leaves entire, linear, acute; in whorls of 8 or 12.

1. **MYRIOPHYL'LUM**, Vaill. Water-Milfoil.

1. **M. spica'tum**, L. *Stamens 8*. Bracts ovate, entire, *shorter than the flowers*. Leaves in whorls of 3 or 4. Flowers greenish, in terminal spikes. Stem very long.—Deep water.

2. **M. verticilla'tum**, L. *Stamens 8*. Leaves finely dissected and whorled as in No. 1. Bracts pectinate-pinnatifid, much longer than the flowers, and the spike therefore leafy. Stem 2-4 feet long.—Stagnant water.

3. **M. heterophyl'lum**, Michx. *Stamens 4*. Lower leaves dissected, in whorls of 4 or 5. Bracts ovate or lanceolate, finely serrate, crowded, the lower ones pinnatifid. Stem stout.—Stagnant or slow water.

2. **HIPPU'RIS**, L. Mare's Tail.

H. vulga'ris, L. A perennial aquatic, with jointed erect stem. —Muddy margins of ponds and streams.

Order XXXVII. **ONAGRA'CEÆ.** (Evening-Primrose F.)

Herbs, with perfect and symmetrical flowers, the parts of the latter in twos or fours. Calyx-tube adherent to the ovary, and usually prolonged above it. Petals and stamens inserted on the calyx. Style 1. Stigmas 2 or 4 or capitate. (See Part I, sections 44-47, for description of a typical plant.)

Synopsis of the Genera.

1. **Circæ'a.** *Petals 2, obcordate. Stamens 2.* Stigma capitate. Fruit bur-like, 1-2-seeded, beset with hooked bristles. Delicate low plants with opposite leaves and very small white flowers in racemes.
2. **Epilo'bium.** *Petals 4. Stamens 8.* Calyx-tube hardly prolonged beyond the ovary. Fruit a linear pod, many-seeded, *the seeds provided with tufts of downy hair.*
3. **Œnothe'ra.** *Petals 4. Stamens 8.* Stigma 4-lobed. *Flowers yellow.* Calyx-tube much prolonged. Pods cylindrical or club-shaped. *Seeds without tufts.*
4. **Ludwig'ia.** *Petals 4, or none. Stamens 4.* Calyx-tube not prolonged. Stigma capitate.

1. CIRCÆ'A, Tourn. ENCHANTER'S NIGHTSHADE.

1. **C. Lutetia'na, L.** Stem 1-2 feet high. Leaves opposite, ovate, slightly toothed. *No bracts under the pedicels.* Fruit roundish, *bristly-hairy, 2-celled.*—Rich woods.

2. **C. alpi'na, L.** Stem *low* and delicate (3-8 inches). Leaves cordate, coarsely toothed. *Minute bracts under the pedicels.* Fruit *club-shaped, soft-hairy, 1-celled.*—Deep low woods.

2. EPILO'BIUM, L. WILLOW-HERB.

1. **E. angustifo'lium, L.** (GREAT WILLOW-HERB.) Stem 3-6 feet high, simple. Leaves lanceolate. Flowers purple, very showy, *in a terminal raceme or spike.* Stigma of 4 long lobes.— Newly-cleared land.

2. **E. palustre, L.,** var. **lineare.** Stem 1-2 feet high, erect, slender, branching above, *hoary-pubescent.* Leaves linear, nearly entire. Flowers *small, corymbed* at the ends of the branches, purplish or white. Petals erect. Stigma club-shaped.—Bogs.

3. **E. molle, Torr.,** is occasionally met with. It differs from No. 2 chiefly in having the leaves crowded and their points more obtuse. The petals are rose-coloured.—Bogs.

4. **E. colora'tum, Muhl.** Stem 1-2 feet high, *nearly smooth.* Leaves *lanceolate* or *ovate-lanceolate.* Flowers small, corymbed. Petals purplish, *deeply notched.*—Extremely common in wet places.

3. ŒNOTHE'RA, L. EVENING PRIMROSE.

1. **Œ. bien'nis, L.** (COMMON EVENING PRIMROSE.) *Stem 2-4 feet high,* hairy. Leaves ovate-lanceolate. Flowers yellow,

odorous, in a leafy spike, opening in the evening or in cloudy weather. Pods oblong, narrowing towards the top.—Waste places.

2. **Œ. pu'mila,** L. (SMALL E.) Stem low, *5-12 inches high*, smooth or nearly so. Leaves lanceolate or oblanceolate. Pods nearly sessile, club-shaped, *4-angled*. Flowers pale yellow, opening in sunshine.—River and lake margins.

3. **Œ. chrysan'tha,** Michx. Distinguished from the preceding by the orange-yellow flowers, and *pedicelled pods*, the latter scarcely wing-angled.—Along the Niagara river.

4. LUDWIG'IA, L. FALSE LOOSESTRIFE.

L. palustris, Ell. (WATER PURSLANE.) Stems creeping in the mud of ditches or river margins, smooth. Leaves opposite, tapering into a slender petiole. Flowers sessile, solitary, usually without petals. Pod 4-sided.

ORDER XXXVIII. **MELASTOMA'CEÆ.** (MELASTOMA F.)

Low herbs with opposite 3-5-ribbed leaves. Calyx-tube adherent to the ovary, the limb 4-cleft. Petals 4, showy, convolute in the bud. Stamens 8, with 1-celled anthers opening by a pore at the apex; these and the petals inserted on the calyx. Style and stigma 1. Pod 4-celled, many-seeded; seeds coiled. The only representative with us is

RHEXIA, L. DEER-GRASS. MEADOW-BEAUTY.

1. **R. Virginica,** L. Stem square, wing-angled. Leaves oval-lanceolate. Petals purple.—Shores of the Muskoka Lakes.

ORDER XXXIX. **LYTHRA'CEÆ.** (LOOSESTRIFE F.)

Herbs, or slightly woody plants, with opposite or whorled entire leaves, without stipules. Calyx enclosing, *but free from*, the ovary. Petals (mostly 5) and stamens (mostly 10) inserted on the calyx. Flowers axillary or whorled. Style 1. Stigma capitate. The only common representative genus with us is

NESÆ'A, Commerson, Juss. SWAMP LOOSESTRIFE.

N. verticilla'ta, H.B.K. Stems curving, 2-6 feet long, 4-6-sided. Leaves lanceolate, mostly whorled. Flowers purple, in the axils of the upper leaves. Calyx bell-shaped, with 5-7 erect teeth, with supplementary projections between them. Stamens 10, exserted, 5 longer than the rest.—Swamps.

ORDER XL. **CUCURBITA′CEÆ.** (GOURD FAMILY.)

Herbs, climbing by tendrils. Flowers monœcious. Calyx-tube adherent to the 1-3-celled ovary. Corolla commonly more or less gamopetalous. Stamens usually 3, united by their tortuous anthers, and often also by the filaments. Leaves alternate, palmately lobed or veined.

Synopsis of the Genera.

1. **Si′cyos.** Flowers greenish-white, small; the staminate corymbed, the pistillate clustered in a head on a long peduncle. *Corolla 5-cleft*, with a spreading border. Style slender; stigmas 3. Ovary 1-celled. Fruit dry and *indehiscent*, prickly, bur-like in appearance.
2. **Echinocys′tis.** Flowers whitish, small; the staminate in long compound racemes, the pistillate in small clusters from the same axils. *Corolla 6-parted.* Stigma broad, almost sessile. Ovary 2-celled, 4-seeded. Fruit fleshy, becoming dry, clothed with weak prickles.

1. SI′CYOS, L. STAR CUCUMBER.

S. angula′tus, L. A clammy-hairy weed in damp yards. Leaves roundish heart-shaped, 5-angled or lobed.

2. ECHINOCYS′TIS, Torr. and Gray. WILD BALSAM-APPLE.

E. loba′ta, Torr. and Gray. Climbing high about dwellings. Leaves deeply and sharply 5-lobed. The oval fruit 2 inches long.

ORDER XLI. **FICOI′DEÆ.** (ICE-PLANT FAMILY.)

A miscellaneous group, embracing plants formerly included in Caryophyllaceæ and Portulacaceæ; differing, however, from true representatives of these in having *partitions in the ovary*. Petals wanting in our genus.

MOLLU′GO, L. CARPET-WEED.

M. verticilla′ta, L. A prostrate much-branched herb, growing in patches. Leaves spathulate, apparently verticillate. Flowers on long axillary pedicels, clustered into a sort of umbel. Sepals 5, white inside. Petals none. Stamens mostly 3. Styles

3. Pod 3-celled, 3-valved, loculicidal, the partitions breaking away from the many-seeded axis.—Mostly in south-western Ontario.

Order XLII. UMBELLIF'ERÆ. (Parsley Family.)

Herbs with small *flowers mostly in compound umbels*. Calyx-tube grown fast to the surface of the ovary ; calyx-teeth minute or none. The 5 petals and 5 stamens inserted on a disk which crowns the ovary. Styles 2. Fruit dry, 2-seeded. Stems hollow. Leaves usually much cut. (See Part I., Chapter VII., for description of a typical flower.)

Synopsis of the Genera.

§ 1. *Seeds flat (not hollow) on the inner face.*

1. **Hydrocot'yle.** *Umbels simple, or one springing from the summit of another,* axillary. Flowers white. Stem slender and creeping. Leaves round-kidney-shaped.

2. **Sanic'ula.** *Umbels irregular* (or compound), the *greenish flowers capitate* in the umbellets. Leaves palmately lobed or parted. Fruit globular, covered with hooked prickles.

(*In the Genera which follow, the umbels are regularly compound.*)

3. **Dau'cus.** Stem bristly. Leaves twice- or thrice-pinnate, or pinnatifid. Bracts of the involucre pinnatifid, very long. Fruit ribbed, the *ribs bristly*.

4. **Heracle'um.** *Stem 3-5 feet high, woolly* and grooved. Leaves 1-2-ternately compound. *Flowers white*, the outer corollas larger than the others. *Fruit wing-margined at the junction of the carpels*, very flat. Carpels 5-ribbed on the back.

5. **Pastina'ca.** *Stem smooth*, grooved. Leaves pinnate. *Flowers yellow, all alike.* Fruit as in No. 4.

6. **Archem'ora.** Stem smooth. Leaves pinnate, of 3-9 rather narrow leaflets. *Flowers white.* Fruit broadly winged, flat, 5-ribbed on the back.

7. **Archangel'ica.** Stem smooth, stout, purple. Leaves 2-3-ternately compound. *Flowers greenish-white.* Fruit smooth, flattish on the back, *double-wing-margined*, each carpel with 3 ribs on the back.

8. **Conioseli'num.** Stem smooth. Leaves finely 2-3-pinnately compound, the petioles inflated. *Flowers white.* Fruit *doubly wing-margined*, and with *3 narrow wings on the back of each carpel.*

9. **Thaspium.** Stem smooth. Leaves 1-2-ternately divided. *Flowers deep yellow.* Fruit not flattened, *10-winged or ribbed.*

10. **Zizia.** Stem slender, smooth and glaucous. Leaves 2-3-ternately compound. *Flowers yellow.* Rays of the umbel long and slender. *Fruit contracted at the junction of the carpels; the carpels narrowly 5-ribbed.*

11. **Cicu'ta.** Stem streaked with purple, stout. Leaves *thrice compound. Flowers white.* Fruit a little contracted at the sides, *the carpels strongly 5-ribbed.*

12. **Sium.** Stem grooved. Leaves simply *pinnate. Flowers white.* Fruit as in No. 10.

13. **Cryptotæ'nia.** Stem smooth. *Leaves 3-foliolate. The umbels with very unequal rays.* Flowers white. Fruit nearly as in Nos. 10 and 11.

§ 2. *Inner face of each seed hollowed lengthwise.*

14. **Osmorrhi'za.** Leaves large, 2-3-ternately compound. Flowers white. Fruit linear-oblong, angled, tapering downwards into a stalk-like base. Ribs of the carpels bristly upwards.

15. **Co'nium.** Leaves large, decompound. Flowers white. Fruit ovate, flattened at the sides, 5-ribbed, *the ribs wavy.*

§ 3. *Inner face of each seed curved inwards at top and bottom.*

16. **Erige'nia.** Stem low and smooth. Leaves 2-3-ternately divided. *Fruit twin.* Carpels nearly kidney-form. Umbels 3-rayed, small. Flowers white.

1. **HYDROCOT'YLE**, Tourn. WATER PENNYWORT.

H. America'na, L. Stem spreading and creeping, very slender. Leaves kidney-shaped, crenate, slightly lobed. Umbels 3-5-flowered, inconspicuous, in the axils of the leaves.—Shady wet places.

2. **SANIC'ULA**, Tourn. SANICLE. BLACK SNAKEROOT.

1. **S. Canaden'sis, L.** *Leaves 3-5-parted. A few staminate flowers* among the perfect ones, and on *very short pedicels.* Styles shorter than the prickles of the fruit.—Low rich woods, not so common as the next.

2. **S. Marilan'dica, L.** *Leaves 5-7-parted.* Staminate flowers numerous, and *on slender pedicels.* Styles long, recurved.—Rich woods.

3. **DAU'CUS**, Tourn. CARROT.

D. Caro'ta, L. (COMMON CARROT.) Found wild occasionally in old fields. In fruit the umbel becomes hollow like a bird's nest.

4. **HERACLE'UM, L.** COW PARSNIP.

H. lana'tum, Michx. Umbels large and flat. Petioles of the leaves spreading and sheathing. Leaves very large; leaflets broadly heart-shaped, deeply lobed. Low wet meadows.

5. **PASTINA'CA,** Tourn. PARSNIP.

P. sati'va, L. (*Peucedanum sativum*, Benth. and Hook., in Macoun's Catalogue.) (COMMON PARSNIP.) Found wild in old fields and along roadsides. Leaflets shining above.

6. **ARCHEMORA,** DC. COWBANE.

A. rig'ida, DC. Calyx 5-toothed. Involucre almost none; involucels of many small bractlets.—Sandy swamps, southwestern Ontario.

7. **ARCHANGEL'ICA,** Hoffm. ARCHANGELICA.

A. atropurpu'rea, Hoffm. (GREAT ANGELICA.) Stem very tall (4-6 feet) and stout, dark purple. Whole plant strongscented. Petioles much inflated at the base.—Marshes and low river-banks.

8. **CONIOSELI'NUM,** Fischer. HEMLOCK-PARSLEY.

C. Canadense, Torr. and Gr. (*Selinum Canadense*, Michx., in Macoun's Catalogue.) Stem 2-4 feet high. Petioles much inflated. Leaflets of the involucels awl-shaped.—Swamps.

9. **THASPIUM,** Nutt. MEADOW-PARSNIP.

T. au'reum, Nutt. Stem 1-2 feet high, angular-furrowed. Leaflets oblong-lanceolate, sharply serrate. *Fruit with 10 winged ridges*, or in var. **apterum** *with 10 ribs.*—Dry or rich woods.

10. **ZIZIA,** DC. ZIZIA.

Z. integer'rima, DC. (*Pimpinella integerrima*, Benth. and Hook., in Macoun's Catalogue.) Stem slender, 1-2 feet high. Involucels none. Plant strong-scented.—Rocky hill-sides.

11. **CICU'TA,** L. WATER-HEMLOCK.

1. **C. macula'ta,** L. (SPOTTED COWBANE. BEAVER POISON.) Stem 3-6 feet high, purplish, smooth. Leaflets *ovate-lanceolate, coarsely serrate*, pointed.—Swamps and low grounds.

2. **C. bulbif'era,** L., is easily distinguished from No. 1 by bearing *clusters of bulblets* in the axils of the upper leaves. The leaflets, also, are *linear.*—Swamps and low grounds.

12. SIUM, L. WATER-PARSNIP.

S. lineare, Michx. (*S. cicutæfolium*, Gmelin, in Macoun's Catalogue.) Stem 2-3 feet high, furrowed. Leaflets varying from linear to oblong, sharply pointed and serrate.—Borders of marshes, usually in the water.

13. CRYPTOTÆ'NIA, DC. HONEWORT.

C. Canadensis, DC. Stem 1-2 feet high, slender. Leaflets large, ovate, doubly serrate. No involucre.—Rich woods and thickets.

14. OSMORRHI'ZA, Raf. SWEET CICELY.

1. **O. longis'tylis**, DC. (SMOOTHER SWEET CICELY.) Stem reddish, nearly smooth. Leaflets sparingly pubescent, short-pointed *Styles slender, nearly as long as the ovary*, recurved.—Rich woods.

2. **O. brevis'tylis**, DC. (HAIRY SWEET CICELY.) Whole plant hairy. Leaflets taper-pointed. *Styles very short, conical.* Rich woods.

15. CO'NIUM, L. POISON HEMLOCK.

C. macula'tum, L. Stem smooth, spotted. Leaflets lanceolate, pinnatifid, pale green, with an offensive odour when bruised. Involucels one-sided. Inner face of the seed marked with a deep groove.—Waste places.

16. ERIGE'NIA, Nutt. HARBINGER-OF-SPRING.

E. bulbo'sa, Nutt. Stem 4-6 inches high, from a tuber deep in the ground, producing 2 leaves, the lower radical. Leaflets much incised. Flowers few.—Alluvial soil.

ORDER XLIII. ARALIA'CEÆ. (GINSENG FAMILY.)

Herbs (with us) differing from the last Order chiefly in having, as a rule, *more than 2 styles*, and the *fruit a drupe*. The umbels, also, are either single, or corymbed, or panicled. Flowers often polygamous. The only Canadian genus is

ARA'LIA, Tourn. GINSENG. WILD SARSAPARILLA.

* *Umbels corymbed or panicled. Petals, stamens, and styles each 5. Fruit black or dark-purple.*

1. **A. racemosa**, L. (SPIKENARD.) *Umbels in a large compound panicle.* Stem 2-3 feet high, widely branching. Leaves

very large and decompound; leaflets ovate-cordate, doubly serrate. Roots aromatic.—Rich woods.

2. **A. his'pida**, Michx. (BRISTLY SARSAPARILLA. WILD ELDER.) Stem 1-2 feet high, *bristly*, leafy, somewhat shrubby at the base. Umbels 2-7, corymbed. Leaves twice-pinnate. Leaflets sharply serrate. Fruit black.—Rocky or sandy woods.

3. **A. nudicau'lis**, L. (WILD SARSAPARILLA.) True stem very short, sending up a naked scape bearing 3 or 4 long-peduncled umbels at the summit, and one long-petioled leaf, ternately divided, and with 5 leaflets on each division. Root horizontal, aromatic.—Rich woods.

* * *Umbel single, on a long peduncle. Styles 2 or 3.*

4. **A. quinquefo'lia**, Decaisne. (GINSENG.) Leaves in a whorl of 3 at the summit of the stem, the latter a foot high. *Leaflets mostly 5, long-stalked.*—Rich woods.

5. **A. trifo'lia**, Decaisne. Stem 4-6 inches high. Leaves in a whorl of 3 at the summit, but *the leaflets usually only 3, and sessile.*—Rich woods.

ORDER XLIV. **CORNA'CEÆ**. (DOGWOOD FAMILY.)

Shrubs or trees (rarely herbs) with simple leaves. Calyx-tube adherent to the 1-2-celled ovary, the limb of the calyx inconspicuous. Petals 4. Stamens 4, ... epigynous. Style 1; stigma flat or capitate. Fruit a 1-2-seeded drupe. Flowers in cymes or in close heads, surrounded by a showy involucre resembling a corolla. The only common Canadian genus is

CORNUS, Tourn. CORNEL. DOGWOOD.

* *Flowers in a close head, surrounded by a showy involucre of 4 white bracts. Fruit red.*

1. **C. Canadensis**, L. (BUNCH-BERRY.) Stem simple, 5 or 6 inches high. Upper leaves crowded and apparently whorled, ovate, the lower scale-like. Leaves of the involucre ovate.—Rich woods.

2. **C. florida**, L. (FLOWERING DOGWOOD.) A small tree, with opposite ovate pointed leaves. Leaves of the involucre notched at the apex.—Rocky woods. South-westward.

' Flowers (white) in flat cymes. No involucre. Fruit blue or white.

3. **C. circina′ta,** L'Her. (ROUND-LEAVED DOGWOOD.) A shrub, 4-6 feet high, with *greenish warty-dotted branches*. Leaves opposite, *broadly oval*, white-woolly beneath. Fruit light blue. —Rich woods.

4. **C. seric′ea,** L. (SILKY CORNEL.) A large shrub, with *purplish branches*. Leaves opposite, narrowly ovate or oblong, silky beneath. Branchlets often rusty. Fruit light blue. Distinguished from No. 3 by the colour of the branches and the much smaller leaves.—Low wet grounds.

5. **C. stolonif′era,** Michx. (RED-OSIER DOGWOOD.) A shrub forming clumps by the production of suckers or stolons, 3-6 feet high. *Branches bright red-purple, smooth.* Leaves opposite, ovate, roughish, whitish beneath. Fruit white or whitish.—Low wet grounds.

6. **C. panicula′ta,** L'Her. (PANICLED CORNEL.) A shrub 4-8 feet high, with erect, gray, and smooth branches. Flowers white, very numerous. Leaves opposite, ovate-lanceolate, taper-pointed. Cymes convex. Fruit white.—Thickets and river-banks.

7. **C. asperifo′lia,** Michx., (ROUGH-LEAVED DOGWOOD) is reported by Macoun as common on Point Pelee. Branches brownish, the branchlets rough-pubescent. Leaves opposite, rather small, oblong or ovate; rough above, downy beneath. Fruit bluish.

8. **C. alternifo′lia,** L. (ALTERNATE-LEAVED CORNEL.) A large shrub or small tree, with *alternate greenish branches streaked with white*. Leaves mostly alternate, oval, acute at each end, crowded at the ends of the branches. Flowers yellowish, in loose cymes. Fruit deep blue, on reddish stalks.—Thickets.

II. GAMOPET′ALOUS DIVISION.

Embracing plants with both calyx and corolla, the latter with the petals united (in however slight a degree).

ORDER XLV. **CAPRIFOLIA′CEÆ.** (HONEYSUCKLE F.)

Shrubs, rarely herbs, with the calyx-tube adherent to the ovary, the corolla borne on the ovary, and the stamens on the tube of the corolla. Leaves opposite and without stipules, but some species of **Vibur'num** have appendages resembling stipules. Fruit a berry, drupe, or pod.

Synopsis of the Genera.

* *Corolla tubular, sometimes 2-lipped. Style slender.*

1. **Linnæ'a.** A trailing or creeping herb, with evergreen oval crenate leaves and slender scape-like peduncles which fork at the top into 2 pedicels, each of which bears a pair of nodding narrowly bell-shaped purplish flowers. Stamens 4, 3 shorter than the others.
2. **Symphoricar'pus.** Upright branching shrubs, with oval entire short-petioled leaves. Flowers in interrupted spikes at the ends of the branches, rose-coloured. Corolla bell-shaped, 4-5-lobed, with as many stamens. Berries large and white, 4-celled, but only 2-seeded.
3. **Lonice'ra.** Upright or twining shrubs, with entire leaves. Corolla funnel-form, more or less irregular, *often with a projection on one side at the base.* Berry several-seeded.
4. **Diervil'la.** Low upright shrubs with ovate pointed serrate leaves. Calyx-tube tapering towards the top, the teeth slender. Flowers light yellow, peduncles mostly 3-flowered. Corolla funnel-form, nearly regular. Pod slender, pointed.
5. **Trios'teum.** Coarse herbs. Lobes of the calyx leaf-like. Flowers brownish-purple, sessile in the axils of the leaves. Corolla bulging at the base. Fruit a 3-seeded orange-coloured drupe.

** *Corolla rotate or urn-shaped, regular, 5-lobed. Flowers white, in broad cymes.*

6. **Sambu'cus.** Upright shrubs with pinnate leaves, the leaflets serrate. Stigmas 3. Fruit purple or red, a juicy berry-like drupe, with 3 seed-like stones.
7. **Vibur'num.** Upright shrubs with simple leaves, and white flowers in compound cymes. Fruit a 1-seeded drupe.

1. LINNÆ'A, Gronov. TWIN-FLOWER.

L. borea'lis, Gronov.—Cool mossy woods and swamps.

2. SYMPHORICAR'PUS, Dill. SNOWBERRY.

S. racemo'sus, Michx. (SNOWBERRY.) Corolla bearded inside. Flowers in a rather loose spike. Var. **pauciflo'rus,** Robbins, is low, diffusely branched, and spreading, with two or three flowers only, in the axils of the uppermost leaves.—Dry rocky hill-sides.

3. LONICE'RA, L. HONEYSUCKLE. WOODBINE.

1. **L. parviflo'ra,** Lam. (*L. glauca,* Hill, in Macoun's Catalogue.) (SMALL HONEYSUCKLE.) *Twining shrub,* 2-4 feet high,

with smooth leaves which are glaucous beneath, the upper ones connate-perfoliate ; corolla yellowish-purple.—Rocky banks.

2. **L. hirsu'ta,** Eaton. (HAIRY HONEYSUCKLE.) Stem *twining high*. Leaves *not glaucous, very large, downy-hairy*, the upper ones connate-perfoliate. Flowers in close whorls; corolla greenish-yellow, clammy-pubescent.—Damp thickets.

3. **L. cilia'ta,** Muhl. (FLY-HONEYSUCKLE.) A branching upright shrub, with *thin oblong-ovate ciliate leaves*. Peduncles axillary, filiform, shorter than the leaves, *each 2-flowered at the top*. Corolla greenish-yellow, *almost spurred at the base*. The two berries separate.—Damp woods.

4. **L. oblongifo'lia,** Muhl. (SWAMP FLY-HONEYSUCKLE). A shrub with upright branches, and *oblong leaves. Peduncles long and slender, 2-flowered*. Corolla deeply 2-lipped. Berries united at the base.—Swamps and low grounds.

4. **DIERVIL'LA,** Tourn. BUSH-HONEYSUCKLE.

D. **trif'ida,** Mœnch.—Rocky woods and clearings.

5. **TRIOS'TEUM,** L. FEVER-WORT.

T. **perfolia'tum,** L. A coarse herb, 2-4 feet high, soft-hairy. Leaves oval, narrowed at the base. Fruit orange-coloured.—Old clearings and thickets.

6. **SAMBU'CUS,** Tourn. ELDER.

1. **S. Canadensis,** L. (COMMON ELDER.) Shrub 5-10 feet high, in clumps. Leaflets 7-10, oblong. Cymes flat. Fruit black-purple.—Open grounds, and along streams.

2. **S. pubens,** Michx., (RED-BERRIED ELDER) may be distinguished from No. 1 by its warty bark, brown pith, 5-7 leaflets. convex or pyramidal cymes, and red berries.—Rocky woods.

7. **VIBUR'NUM,** L. ARROW-WOOD. LAURESTINUS.

1. **V. Lenta'go,** L. (SWEET VIBURNUM. SHEEP-BERRY.) A small tree, with *ovate finely-serrate pointed leaves*, with long and margined petioles. *Cyme sessile*. Fruit black.—Along streams.

2. **V. nudum,** L. (WITHE-ROD.) A smooth shrub with tall straight stems. Leaves thickish, entire or wavy-toothed, dotted beneath. *Cymes with short peduncles*. Fruit black.—Cold swamps.

3. **V. pubes'cens,** Pursh. (DOWNY ARROW-WOOD.) A straggling shrub, not more than 4 feet high, with small ovate coarsely serrate leaves, *the lower surface soft-downy.* Cymes small. Fruit oblong, dark-purple.—Rocky places.

4. **V. acerifo'lium,** L. (MAPLE-LEAVED A. DOCKMACKIE.) A shrub 3-6 feet high, with greenish bark. Leaves 3-lobed, 3-ribbed, soft-downy beneath. Stipular appendages bristle-shaped. Cymes small, on long peduncles. Fruit red, becoming black.—Thickets and river-banks.

5. **V. Op'ulus,** L. (CRANBERRY-TREE.) An upright shrub, 5-10 feet high, with strongly 3-lobed leaves, broader than long, the lobes spreading and pointed. Cymes peduncled. Marginal flowers of the cyme very large and neutral. Stipular appendages conspicuous. Fruit red, pleasantly acid.—Low grounds.

6. **V. lantanoi'des,** Michx. (HOBBLE-BUSH.) A straggling shrub with reclining branches. Leaves large, round-ovate, heart-shaped at the base, serrate, many-veined, the veins underneath and the stalks and branchlets very rusty-scurfy. Stipular appendages conspicuous. Cymes sessile, very broad and flat, with very conspicuous neutral flowers on the margin.—Moist woods.

ORDER XLVI. **RUBIA'CEÆ.** (MADDER FAMILY.)

Herbs or shrubs, chiefly distinguished from the preceding Order by the presence of stipules between the opposite entire leaves, or by the leaves being in whorls without stipules. Calyx superior. Stamens alternate with the (mostly 4) lobes of the corolla, and inserted on its tube. Ovary 2-4-celled.

Synopsis of the Genera.

1. **Ga'lium.** *Leaves in whorls. Slender weak herbs with square stems.* Calyx-teeth inconspicuous. Corolla 4-parted, wheel-shaped. Styles 2. Fruit twin, separating into two 1-seeded carpels.
2. **Cephalan'thus.** *Leaves opposite. Shrubs with the flowers in a globular peduncled head.* Lobes of calyx and corolla each 4. Style very slender, much protruded. Stigma capitate.
3. **Mitchel'la.** *Leaves opposite. Shining trailing evergreen herbs,* with flowers in pairs, *the ovaries united.* Lobes of calyx and corolla each 4, the corolla bearded inside. Style 1. Stigmas 4. Fruit a red 2-eyed berry.

4. **Housto'nia.** *Leaves opposite. Low and slender erect herbs*, with the flowers in small terminal clusters. Lobes of calyx and corolla each 4. Style 1. Stigmas 2.

1. GALIUM, L. BEDSTRAW. CLEAVERS.

1. **G. Apari'ne, L.** (CLEAVERS. GOOSE-GRASS.) *Leaves about 8 in a whorl*, lanceolate, rough-margined. Peduncles 1-2-flowered, axillary. *Fruit covered with hooked prickles.*—Low grounds.

2. **G. triflo'rum,** Michx. (SWEET-SCENTED BEDSTRAW.) *Leaves chiefly 6 in a whorl*, elliptical-lanceolate, bristle-pointed. Peduncles 3-flowered, *terminating the branches. Fruit covered with hooked prickles.*—Woods.

3. **G. pilo'sum,** Ait. Leaves in whorls of 4, *hairy, oval.* Peduncles twice- or thrice-forked.—Southwestern Ontario.

4. **G. lanceola'tum,** Torr. (WILD LIQUORICE.) *Leaves all in whorls of 4 each*, lanceolate, *tapering at the apex*, more or less 3-nerved. Peduncles mostly once-forked. Flowers few or several, remote. Fruit covered with hooked prickles.

5. **G. circæ'zans,** Michx., is similar to No. 3, but the leaves are *obtuse* instead of tapering.—Woods.

6. **G. asprel'lum,** Michx. (ROUGH BEDSTRAW.) Leaves in whorls of 6, or 4 or 5 on the branchlets, elliptical-lanceolate, very rough on the edges and midrib. Stem weak, 3-5 feet high, leaning upon and clinging to bushes by its rough edges. *Flowers numerous in panicled clusters.* Fruit not rough.—Thickets.

7. **G. trif'idum, L.** (SMALL BEDSTRAW.) Leaves in whorls of 4-6. Stem 6-18 inches high, roughened on the edges, as are the leaves usually. *Flowers few, not panicled. Parts of the flowers generally in threes.* Fruit smooth. Var. **latifolium,** Torr., is easily known by its broad leaves and widely branching stems.—Low grounds and swamps.

8. **G. borea'le, L.** (NORTHERN BEDSTRAW.) *Leaves in whorls of 4*, linear-lanceolate, 3-nerved. Flowers very numerous, crowded in a narrow and compact terminal panicle. Stem erect and rigid, 1-3 feet high.—Rocky thickets and river-banks.

3. CEPHALANTHUS, L. BUTTON BUSH.

C. occidenta'lis, L. A smooth shrub growing in swamps, with ovate petioled pointed leaves, which are opposite or in whorls of 3. Easily recognized by the globular head of white flowers.

4. MITCHEL′LA, L. PARTRIDGE BERRY.

M. repens, L.—Common in dry woods. Leaves round-ovate, shining, sometimes with whitish lines.

5. HOUSTO′NIA, L. HOUSTONIA.

H. purpu′rea, L. Stems tufted, 3-6 inches high. Leaves varying from roundish-ovate to lanceolate, 3-5-ribbed, sessile.—Woodlands.

ORDER XLVII. **VALERIANA′CEÆ.** (VALERIAN F.)

Herbs with opposite exstipulate leaves, and small cymose flowers. Calyx-tube adherent to the ovary, the latter 3-celled, but *only one of these fertile*. Stamens 1-3, *fewer than the lobes of the corolla*. Style slender. Stigmas 1-3. The only common genus is

VALERIA′NA, Tourn. VALERIAN.

1. **V. sylvat′ica**, Richards. (*V. dioica*, var. **uliginosa**, Torr. and Gray, in Macoun's Catalogue.) Not uncommon in cedar-swamps. *Root fibrous. Calyx-limb consisting of several bristles rolled inwards in the flower, but expanding in fruit.* Corolla gibbous at the base. Stamens 3. Root-leaves ovate or oblong, entire; stem-leaves pinnate, leaflets 5-11. Stem erect, striate, 1-2 feet high.

2. **V. ed′ulis**, Nutt. *Root spindle-shaped, large.* Flowers in a long and narrow interrupted panicle, nearly diœcious. Stem-leaves deeply pinnatifid.—Low grounds, western Ontario.

ORDER XLVIII. **DIPSA′CEÆ.** (TEASEL FAMILY.)

Herbs with the flowers in heads, surrounded by a many-leaved involucre, as in the next Family, *but the stamens are distinct.* Leaves opposite. Represented in Canada by the genus

DIP′SACUS, Tourn. TEASEL.

D. sylves′tris, Mill. (WILD TEASEL.) A stout coarse prickly plant, not unlike a thistle in appearance. Flowers in oblong very dense heads, bluish. Corolla 4-cleft. Stamens 4, on the corolla. Bracts among the flowers terminating in a long awn. Leaves generally connate.—Roadsides and ditches. Rather common in the Niagara district, but found also elsewhere.

Order XLIX. COMPOSITÆ. (Composite Family.)

Flowers in a dense head on a common receptacle, and surrounded by an involucre. Calyx-tube adherent to the ovary, its limb either obsolete or forming a pappus of few or many bristles or chaffy scales. Corolla either tubular or with one side much prolonged (strap-shaped or ligulate). Stamens usually 5, on the tube of the corolla, their anthers united (syngenesious). Style 2-cleft. (See Part I., sections 60-62, for examination of a typical flower.)

The heads of flowers present some variety of structure. *All* the flowers of a head may be tubular; or only the central ones or *disk-flowers*, as they are then called, may be tubular, whilst those around the margin, then known as *ray-flowers*, are ligulate or strap-shaped. Or again, *all* the flowers may be strap-shaped. It is not unusual also to find a mixture of perfect and imperfect flowers in the same head.

The bracts which are often found growing on the common receptacle among the florets are known as the *chaff*. When these bracts are entirely absent the receptacle is said to be *naked*. The leaves of the involucre are called its *scales*.

Artificial Synopsis of the Genera.

Suborder I. **TUBULIF'LORÆ.**

Heads either altogether without strap-shaped corollas, or the latter, if present, forming only the outer circle (the *ray*). Ray-flowers, when present, *always without stamens*, and often without a pistil also.

A. **Ray-flowers entirely absent.**

* Scales of the involucre in many rows, *bristly-pointed, or fringed.*

← *Florets all perfect.*

1. **Cir'sium.** *Leaves and scales of the involucre prickly.* Pappus of long plumose bristles. Receptacle with long soft bristles among the florets. Flowers reddish-purple.

2. **Onopor'don.** *Leaves and scales of the involucre prickly.* Heads much as in Cirsium, but the *receptacle naked*, and deeply honeycombed. Pappus of long bristles, *not plumose.* Stem winged by the decurrent bases of the leaves. Flowers purple.

3. **Lap'pa.** *Leaves not prickly, but the scales of the globular involucre tipped with hooked bristles.* Pappus of many short rough bristles. Receptacle bristly. Flowers purple.

← ← *Marginal florets sterile, and their corollas much larger than the others, forming a kind of false ray.*

4. **Centaure'a.** *Leaves not prickly. Scales of the involucre fringed.* Pappus very short. Receptacle bristly. Flowers blue.

← ← ← *Sterile and fertile florets in separate heads,* i.e., *monœcious. Fruit a completely closed involucre (usually bristly) containing only one or two florets, these heads sessile in the axils of the bracts or upper leaves. Sterile heads with more numerous florets in flattish involucres, and forming racemes or spikes. Pappus none.*

5. **Xan'thium.** Fertile florets only 2 together in burs with hooked prickles, clustered in the axils. Sterile heads in short spikes above them, the scales of their involucres in one row only, *but not united together.*

6. **Ambro'sia.** Fertile florets single, in a closed involucre armed with a few spines at the top. Sterile heads in racemes or spikes above, the scales of their involucres in a single row and *united into a cup.*

* * Scales of the involucre without bristles of any kind.

+ Marginal florets without stamens.

++ *Pappus none or minute. Receptacle naked. Very strong-scented herbs.*

7. **Tanace'tum.** Flowers yellow, in numerous corymbed heads. Scales of the involucre dry, imbricated. Pappus 5-lobed. Leaves dissected.

8. **Artemis'ia.** Flowers yellowish or dull purplish, in numerous small heads which are panicled or racemed. Scales of the involucre with dry and scarious margins, imbricated. Hoary herbs.

++ ++ *Pappus of all the florets bristly. Receptacle naked.*

9. **Erechthi'tes.** Flowers whitish. *Scales of the involucre in a single row, linear, with a few bractlets at the base.* Corolla of the marginal florets very slender. Pappus copious, of fine soft white hairs. Heads corymbed. Erect and coarse herbs.

10. **Gnapha'lium.** Flowers whitish or yellowish. Scales of the involucre yellowish-white, in many rows, dry and scarious, woolly at the base. Outer corollas slender. Pappus a single row of fine rough bristles. Flocculent-woolly herbs.

11. **Antenna'ria.** Very much like Gnaphalium in appearance, being white-woolly, *but the heads are usually diœcious,* and the bristles of the pappus thicker in the sterile florets.

+ + **All the florets in the head perfect.**

11. **Antenna'ria,** with diœcious heads, may be looked for here. See previous paragraph.

    Bidens. One or two species have no rays. See No. 25.

    Sene'cio. One species is without rays. See No. 14.

12. **Lia'tris.** Flowers handsome, rose-purple. Receptacle naked. Pappus of long and slender bristles, plumose or rough. Achenes slender, 10 ribbed. Lobes of the corolla slender. Stem wand-like, leafy *Leaves narrow or grass-like.*

13. **Eupato'rium.** Flowers white or purple. Receptacle naked. Pappus of slender hair-like bristles, smooth or nearly so. Achenes 5-angled. Heads in corymbs. Leaves whorled, or connate, or opposite.

B. **Rays or strap-shaped corollas round the margin of the head.**

   * *Pappus of hair-like bristles. Receptacle naked.*

14. **Sene'cio.** Rays yellow, or in one species *none*. Scales of the involucre in a single row, or with a few bractlets at the base. Pappus very fine and soft. Heads corymbose. Leaves alternate.

15. **In'ula.** Rays yellow, numerous, very narrow, in a single row. Outer scales of the involucre leaf-like. *Anthers with two tails at the base.* Stout plants, with large alternate leaves which are woolly beneath.

16. **Solida'go.** Rays yellow, few, as are also the disk-florets. Involucre oblong, scales of unequal lengths, appressed. Achenes many-ribbed. Heads small, in compound racemes, or corymbs. Stems usually wand-like. Leaves alternate.

17. **Nardos'mia.** Rays whitish or purplish. Heads in a corymb, fragrant. Scales of the involucre in a single row. Heads somewhat diœcious, the staminate with one row of pistillate ray-flowers, the pistillate with ray-flowers in many rows. *Woolly herbs,* with large leaves, *all radical,* and sheathing scaly bracts on the scape.

18. **Aster.** Rays white, purple, or blue, *never yellow,* but the *disk* generally yellow. Pappus a *single row* of numerous fine roughish bristles. Achenes flattish. Heads corymbed or racemose. Flowering in late summer.

19. **Erig'eron.** Rays and disk as in Aster, *but the rays very narrow, and usually in more than one row.* Scales of the involucre in one or two rows, nearly of equal length. Pappus of long bristles *with shorter ones intermixed.* Heads corymbed or solitary. Leaves generally sessile.

20. **Diplopap'pus.** Rays white, long. Disk-florets yellow. Scales of the involucre 1-nerved. *Pappus double,* the outer row of short stiff bristles. Heads small. corymbed.

* * *Pappus not of hair-like bristles, but either altogether wanting or consisting of a few chaffy scales or teeth, or only a minute crown.*

+ Receptacle naked.

21. **Hele'nium.** Rays yellow, wedge-shaped, 3-5-cleft at the summit. Scales of the involucre reflexed, awl-shaped. Pappus of 5-8 chaffy scales, 1-nerved, the nerve usually extending into a point. Leaves alternate, decurrent on the angled stem. Heads corymbed, showy.

22. **Leucan'themum.** Rays white; disk yellow. Disk-corollas with a flattened tube. *Pappus none.* Heads single.

+ + Receptacle chaffy.

23. **Maru'ta.** Rays white, *soon reflexed;* disk yellow. *Ray-florets neutral.* Pappus none. Receptacle conical, more or less chaffy. Herbs with strong odour.

24. **Rudbeck'ia.** Rays yellow, usually long; *disk dark-purple, or in one species greenish-yellow.* Scales of the involucre leaf-like. Receptacle conical. Pappus none, or only a minute crown. Ray-florets neutral.

25. **Helian'thus.** Rays yellow, neutral. Receptacle flattish or convex. Chaff persistent, and *embracing the 4-sided achenes.* Pappus deciduous, of 2 thin scales. Stout coarse herbs.

26. **Bidens.** Rays yellow, few; but 2 species are without rays. Scales of the involucre in 2 rows, the outer large and leaf-like. Ray-florets neutral. *Achenes crowned with 2 or more stiff awns which are barbed backward.*

27. **Heliop'sis.** Rays yellow, 10 or more, pistillate. Scales of the involucre in 2 or 3 rows, the outer leaf-like. Receptacle conical; chaff linear. Achenia smooth, 4-angled. Pappus none.

28. **Achille'a.** Rays white (occasionally pinkish), few. Receptacle flattish. Pappus none. Achenes margined. Heads small, in flat corymbs. Leaves very finely dissected.

29. **Polym'nia.** Rays whitish-yellow, wedge-form, shorter than the involucre, few in number. Scales of the involucre in 2 rows, the outer leaf-like, the inner small, and partly clasping the achenes. Pappus none. Coarse clammy herbs with an unpleasant odour.

30. **Sil'phium.** Easily known by its stout square stem, and the upper connate leaves forming a sort of cup. Flowers yellow. Achenes broad and flat.

SUBORDER II. **LIGULIFLORÆ.**

Corolla strap-shaped in all the florets of the head. All the florets perfect. Herbs with milky juice, and alternate leaves.

31. **Cyn'thia.** Flowers yellow. Pappus double, the outer short, of many minute chaffy scales, the inner of many long capillary bristles. Low perennials with single showy heads on scapes.

32. **Lamp'sana.** Flowers yellow, 8-12 in a head. Scales of the involucre 8, in a single row. *Pappus none.* Stem slender. Heads small, in loose panicles.

33. **Ciclio'rium.** Flowers bright blue, showy. Scales of the involucre in 2 rows, the outer of 5 short scales, the inner of 8-10 scales. *Pappus chaffy.* Heads sessile, 2 or 3 together.

34. **Leon'todon.** Flowers yellow. Involucre with bractlets at the base. *Pappus of plumose bristles*, these broader at the base. Heads borne on branching scapes. Leaves radical.

35. **Hiera'cium.** Flowers yellow. Scales of the involucre more or less imbricated. *Pappus a single row of tawny hair-like rough bristles.* Heads corymbose.

36. **Nab'alus.** Flowers yellowish- or greenish-white, often tinged with purple; heads nodding. Involucre of 5-14 scales in a single row, with a few bractlets below. *Pappus copious, of brownish or yellowish rough bristles.*

37. **Tarax'acum.** Flowers yellow, *on slender naked hollow scapes.* Achenes prolonged into a slender thread-like beak. Leaves all radical. (See Part I., Chapter viii.)

38. **Lactu'ca.** Flowers pale yellow or purplish. Florets few (about 20) in the head. Scales of the involucre in 2 or more rows of unequal length. *Achenes with long thread-form beaks*, and a pappus of very soft white bristles. Heads numerous, panicled. Tall smooth herbs with runcinate leaves.

39. **Mulge'dium.** Flowers chiefly blue. Structure of the heads and general aspect of the plant as in Lactuca, *but the beak of the achenes short and thick, and the pappus tawny.* Heads in a dense panicle.

40. **Son'chus.** Flowers pale yellow. Heads many-flowered, enlarging at the base. Achenia without beaks. Pappus very soft and white. Tall glaucous herbs with *spiny-toothed leaves.*

41. **Tragopo'gon.** Flowers yellow. Heads large. Involucre of about 12 lanceolate rather fleshy scales in one row, somewhat united at the base. Achenes with long tapering beaks. Pappus of plumose bristles, 5 of these longer and naked at the summit. Leaves entire, straight-veined, clasping.

1. CIR'SIUM, Tourn. COMMON THISTLE.

1. **C. lanceola'tum,** Scop. (*Cnicus lanceolatus*, Hoffm., in Macoun's Catalogue.) (COMMON THISTLE.) *All the scales of the involucre prickly-pointed.* Leaves decurrent, pinnatifid, the lobes prickly-pointed, rough above, woolly with webby hairs beneath. —Fields and roadsides everywhere.

2. **C. dis'color**, Spreng. (*Cnicus altissimus*, Willd., var. **discolor**, Gray, in Macoun's Catalogue.) The *inner* scales of the involucre not prickly. Stem grooved. Leaves prickly, green above, *white-woolly beneath*. Flowers pale purple. Whole plant with a whitish aspect.—Dry thickets.

3. **C. mu'ticum**, Michx. (*Cnicus muticus*, Pursh, in Macoun's Catalogue.) (SWAMP THISTLE.) *Scales of the webby involucre hardly prickly*, and not spreading. *Stem very tall*, and smoothish, and sparingly leafy. *Heads single or few.*—Swamps and low woods.

4. **C. arvense**, Scop. (*Cnicus arvensis*, Pursh, in Macoun's Catalogue.) (CANADA THISTLE.) Scales of the involucre with reflexed points. Leaves prickly, smooth both sides, or slightly woolly beneath. Roots extensively creeping. Heads small and numerous.—Fields and roadsides.

2. ONOPOR'DON, Vaill. SCOTCH THISTLE.

O. acan'thium, L. A coarse branching herb, 2-4 feet high, with woolly stem and leaves. Bristles of the pappus united at the base into a ring.—Roadsides and old fields; not common.

3. LAP'PA, Tourn. BURDOCK.

L. officina'lis, All., var. **major**, Gray. (*Arctium Lappa*, L., in Macoun's Catalogue.) A coarse plant with very large cordate petioled leaves, and numerous small globular heads of purple flowers. The involucre forms a bur which clings to one's clothing, or to the hair of animals.—Near dwellings, mostly in manured soil.

4. CENTAURE'A, L. STAR-THISTLE.

C. Cy'anus, L. (BLUE-BOTTLE.) An old garden plant, found occasionally along roadsides. False rays very large. Scales of the involucre fringed. Leaves linear, entire or nearly so. Stem erect. Heads single at the ends of the branches.

5. XAN'THIUM, Tourn. CLOTBUR.

1. **X. struma'rium**, L., var. **echina'tum**, Gray. (*X. Canadense*, ? ill, var. *echinatum*, Gray, in Macoun's Catalogue.) (COMMON COCKLEBUR.) Stem rough, not prickly or spiny.

Leaves broadly triangular, and somewhat heart-shaped, long-petioled. Fruit a hard 2-celled bur, nearly an inch long, clothed with stiff hooked prickles, the two beaks of the fruit long and usually incurved.—Low river banks.

2. **X. spino'sum.** (SPINY CLOTBUR.) Stem armed with conspicuous straw-coloured triple slender spines, at the bases of the lanceolate short-petioled leaves, the latter white-woolly beneath. —Town of Dundas, Ontario; the seeds having been brought in wool from South America.

6. AMBRO'SIA, Tourn. RAGWEED.

1. **A. artemisiæfo'lia,** L. (HOG-WEED.) Stem erect, 1-3 feet high, branching, hairy. *Leaves twice-pinnatifid,* the lobes linear, paler beneath.—Waste places everywhere, but not so common northward.

2. **A. trif'ida,** L. (GREAT RAGWEED) is found in low grounds in the south-west of Ontario; also at Montreal and Ottawa. Stem stouter than No. 1, 2-4 feet high. *Leaves opposite, deeply 3-lobed,* the lobes oval-lanceolate and serrate.

7. TANACE'TUM, L. TANSY.

T. vulga're, L. (COMMON TANSY.) A very strong-scented herb, 2-4 feet high, smooth. Leaves twice-pinnate, the lobes serrate, as are also the wings of the petiole. Heads densely corymbed. Var. **crispum,** DC., is easily distinguished by its crisper and more incised leaves.—Old gardens and roadsides near dwellings.

8. ARTEMIS'IA, L. WORMWOOD.

1. **A. Canadensis,** Michx. Stem smooth or sometimes hoary with silky down, erect, usually brownish. Lower leaves twice-pinnatifid, the lobes linear.—Shores of the Great Lakes.

2. **A. vulga'ris,** L. (COMMON MUGWORT.) Stem tall, and branching above. *Leaves green and smooth above,* white-woolly beneath, pinnatifid, the lobes linear-lanceolate. Heads small, erect, in panicles. Flowers purplish.—Old fields near dwellings.

3. **A. Absin'thium,** L. (COMMON WORMWOOD.) Somewhat shrubby. Whole plant silky-hoary. Stem angular, branched, the branches with drooping extremities. Leaves 2-3-pinnately-

divided, the lobes lanceolate. Heads nodding.—Escaped from gardens in some places.

9. **ERECHTHITES**, Raf. Fireweed

E. hieracifo'lia, Raf. Stem tall, grooved. Leaves sessile, lanceolate, cut-toothed, upper ones clasping.—Common in places recently over-run by fire.

10. **GNAPHA'LIUM**, L. Cudweed.

1. **G. decur'rens**, Ives. (Everlasting). Stem erect, 2 feet high, *clammy-pubescent*, white-woolly on the branches. Heads corymbed. Leaves linear-lanceolate, partly clasping, *decurrent*. —Fields and hillsides.

2. **G. polyceph'alum**, Michx. (Common Everlasting.) Stem erect, 1-2 feet high, white-woolly. Heads corymbed. Leaves lanceolate, tapering at the base, *not decurrent*.—Old pastures and woods.

3. **G. uligino'sum**, L. (Low Cudweed.) Stem spreading, 3-6 inches high, white-woolly. Leaves linear. Heads small *in crowded terminal clusters* subtended by leaves.—Low ground.

11. **ANTENNA'RIA**, Gærtn. Everlasting.

1. **A. margarita'cea**, R. Brown. (Pearly Everlasting.) (*Anaphalis margaritacea*, Benth. and Hook., in Macoun's Catalogue.) Stem in clusters, downy. Leaves linear-lanceolate, taper-pointed, sessile. Scales of the involucre pearly-white. Heads in corymbs.—Along fences and in open woods.

2. **A. plantaginifo'lia**, Hook. (Plantain-leaved E.) *Stem scape-like, 4-6 inches high*. Radical leaves spathulate or obovate ; stem-leaves few, linear. Heads small, in a crowded corymb. Involucre white or purplish.—Old pastures and woods.

12. **LIA'TRIS**, Schreb. Blazing-Star.

1. **L. cylindra'cea**, Michx. Stem wand-like, 6-18 inches high. Leaves linear, rigid, generally 1-nerved. Heads few, cylindrical. Sandy fields and thickets.

2. **L. spica'ta**, Willd. Stem stout and rigid, 2-5 feet high, very leafy. Leaves linear, erect, the lowest 3-5-nerved. Heads crowded in a long spike.—Low grounds, south-western Ontario.

13. EUPATO'RIUM, Tourn. THOR'UGHWORT.

1. **E. purpu'reum.** L. (JOE-PYE WEED. TRUMPET-WEED.) Stem tall and simple. *Leaves petioled, 3-6 in a whorl. Flowers purplish or flesh-coloured.* Heads in dense corymbs.—Low grounds.

2. **E. perfolia'tum,** L. (BONESET.) Stem short, hairy. *Leaves rugose, connate-perfoliate, tapering.* Flowers whitish. Corymbs very large.—Low grounds.

3. **E. ageratoi'des,** L. (WHITE SNAKE-ROOT.) Stem very smooth, commonly branching, 2-3 feet high. *Leaves opposite, petioled, broadly ovate, pointed, coarsely serrate.* Flowers white, in corymbs.—Low rich woods.

14. SENE'CIO, L. GROUNDSEL.

1. **S. vulga'ris,** L. (COMMON GROUNDSEL.) *Ray-florets wanting.* Stem low, branching. Leaves pinnatifid and toothed, clasping. Flowers yellow, terminal.—Cultivated and waste grounds.

2. **S. au'reus,** L. (GOLDEN RAGWORT. SQUAW-WEED.) *Rays 8-12.* Stem smooth, or woolly when young, 1-2 feet high. Root-leaves simple, rounded, usually cordate, crenately tooth d, long-petioled. Stem-leaves sessile, lanceolate, deeply pinnatifid. Heads in a corymb nearly like an umbel.—Swamps, and often in gardens.

15. IN'ULA, L. ELECAMPANE.

1. **Hele'nium,** L. (COMMON ELECAMPANE.) Stem stout, 2-5 feet high. Root-leaves very large, ovate, petioled. Stem-leaves clasping. Rays numerous, narrow.—Roadsides.

16. SOLIDA'GO, L. GOLDEN-ROD.

* *Heads clustered in the axils of the feather-veined leaves.*

1. **S. squarro'sa,** Muhl. Stem stout, 2-5 feet high, simple, hairy above. *Scales of the involucre with reflexed herbaceous tips.* Leaves large, oblong, serrate, veiny; the lower tapering into a long-winged petiole, the upper sessile and entire. Heads in racemose clusters, *the whole forming a dense, leafy, interrupted, compound spike.*—Rocky woods.

2. **S. bi'color,** L. Stem hoary-pubescent, usually simple. Leaves oval-lanceolate, acute at both ends; the lower oval and

tapering into a petiole, serrate. Heads in short racemes in the upper axils, the whole forming an interrupted spike or compound raceme. *Ray-florets whitish.* The variety **con'color** has *yellow* rays.—Dry banks and thickets.

3. **S. latifo'lia, L.** Stem smooth, not angled, zigzag, 1-3 feet high. Leaves broadly ovate or oval, strongly and sharply serrate, pointed at both ends. Heads in very short axillary clusters.— Cool woods.

4. **S. cæ'sia, L.**, var. **axilla'ris**, Gray. Stem smooth, angled, glaucous, slender, usually branching above. Leaves smooth, lanceolate, pointed, serrate, sessile. Heads in very short clusters in the axils of the leaves.—Rich woods and hillsides.

** *Racemes terminal, erect, loosely thyrsoid, not one-sided. Leaves feather-veined.*

5. **S. Virga-aurea, L.**, var. **hu'milis**, Gray. (*S. humilis*, Pursh, in Macoun's Catalogue.) Stem low, 3-6 inches high, usually smooth; the heads, peduncles, etc., mostly glutinous. Leaves lanceolate or oblanceolate, serrate or entire, the radical ones petiolate, obtuse, and serrate at the apex.—Rocky banks; not common.

*** *Heads in a compound corymb terminating the simple stem, not at all racemose.*

6. **S. Ohioen'sis**, Riddell. Very smooth throughout. Stem slender, reddish, leafy. Radical leaves very long (often a foot), slightly serrate towards the apex, tapering into long margined petioles; stem-leaves oblong-lanceolate, entire, sessile.—Wet grassy shores of Red Bay, Lake Huron.

**** *Heads in one-sided racemes, spreading or recurved.*

7. **S. argu'ta**, Ait. (*S. juncea*, Ait., in Macoun's Catalogue.) Whole plant smooth, 1-4 feet high, rigid, branching above. Lower leaves oval or elliptical-lanceolate, serrate with spreading teeth, pointed, tapering into winged and ciliate petioles; upper ones lanceolate. Racemes very dense, naked, at length elongated and recurved. The variety **juncea** has narrower and less serrate leaves.—Woods and banks.

8. **S. Muhlenber'gii**, Torr. and Gr. (*S. arguta*, Ait., in Macoun's Catalogue.) Stem smooth, angled or furrowed. Leaves

large and thin, ovate; the upper elliptical-lanceolate. Racemes much shorter and looser than in No. 7, and the rays much larger. —Moist woods and thickets.

9. **S. altis'sima,** L. (*S. rugosa*, Mill, in Macoun's Catalogue.) Stem rough-hairy, *less than a foot high*. Leaves ovate-lanceolate or oblong, coarsely serrate, veiny, *often rugose*. Racemes panicled, spreading.—Borders of fields and copses.

10. **S. neglecta,** Torr. and Gr. Stem smooth, 2-3 feet high, stout. Leaves thickish, smooth both sides, the upper oblong-lanceolate, nearly entire, the lower ovate-lanceolate or oblong, sharply serrate, tapering into a petiole. Heads rather large. Racemes short and dense, at first erect and scarcely one-sided, at length spreading.—Swamps.

11. **S. nemora'lis,** Ait. Stem minutely and closely hoary pubescent, simple or corymbed. Leaves more or less hoary, slightly 3-nerved, obscurely serrate or entire; the lower oblanceolate, somewhat crenate, and tapering into a petiole. Racemes numerous, dense, at length recurved, forming a large panicle.— Dry fields.

* * * * * *Racemes one-sided and recurved, and the leaves plainly 3-ribbed.*

12. **S. Canadensis,** L. Stem rough-hairy, tall and stout. Leaves lanceolate, serrate, pubescent beneath, rough above. Panicle exceedingly large.—Very common along fences and in moist thickets.

13. **S. sero'tina,** Ait. (*S. serotina*, Ait., var. *gigantea*, Gray, in Macoun's Catalogue.) Stem very smooth, tall and stout. Leaves lanceolate, serrate, the veins beneath pubescent. Panicle pyramidal, of many curved racemes.—Low thickets and meadows.

14. **S. gigante'a,** Ait. (*S. serotina*, Ait., in Macoun's Catalogue.) Stem smooth, stout. Leaves lanceolate, taper-pointed, sharply serrate, except at the base, smooth both sides, *rough-ciliate*. Panicle large, *pubescent.*—Open thickets and meadows.

* * * * * * *Inflorescence a flat-topped corymb.*

15. **S. lanceola'ta,** L. Stem pubescent above, much branched. Leaves linear-lanceolate, the nerves (3-5) and margins rough-pubescent. Heads in dense corymbed clusters, giving a decidedly characteristic aspect to this species.—Low river-margins.

COMPOSITÆ. 75

17. NARDOS'MIA, Cass. SWEET COLTSFOOT.

N. palma'ta, Hook. Leaves rounded, somewhat kidney-shaped, palmately 5-7-lobed, the lobes toothed and cut.—Cedar swamps and bogs.

18. ASTER, L. STARWORT. ASTER.

* *Leaves, at least the lower ones, heart-shaped and petioled.*

1. **A. corymbo'sus,** Ait. Rays white or nearly so. Heads in corymbs. Stem slender, 1-2 feet high, zigzag. Leaves thin, smoothish, sharp-pointed, coarsely serrate, all the lower ones on slender naked petioles.—Woodlands.

2. **A. macrophyl'lus,** L. Rays white or bluish. Stem stout, 2-3 feet high. Leaves thickish, rough, finely serrate, the lower long-petioled. Heads in closer corymbs than in No. 1.—Woodlands.

3. **A. azu'reus,** Lindl. Rays bright blue. Heads racemed or panicled. Stem roughish, erect, racemose-compound above. Leaves entire or nearly so, rough ; the lower ovate-lanceolate, on long petioles ; the upper lanceolate or linear, sessile. The latest flowering of our Asters.—Dry soil.

4. **A. undula'tus,** L. Rays bright blue. Heads racemed or panicled. Stem hoary with close pubescence, spreading. Leaves with somewhat wavy margins, ovate or ovate-lanceolate, roughish above, downy beneath ; the lowest cordate, on margined petioles ; the upper with *winged short petioles* clasping at the base, or sessile.—Dry woods.

5. **A. cordifo'lius,** L. Rays pale blue or nearly white. Heads *small*, profuse, panicled. Stem much branched. Leaves thin, sharply serrate, the lower on slender ciliate petioles.—Woods and along fences.

6. **A. sagittifo'lius,** Willd. Rays pale blue or purple. Heads *small*, in dense compound racemes or panicles. Stem smooth or nearly so, erect, with ascending branches. Leaves ovate-lanceolate, pointed, pubescent, the lowest on long margined petioles, the upper contracted into a winged petiole, or lanceolate or linear. —Thickets and along fences.

* * *Upper leaves all sessile or clasping by a heart-shaped base ; lower ones not heart-shaped.*

7. **A. lævis,** L. *Rays large,* purple or blue. Very smooth throughout. Heads in a close panicle. Leaves lanceolate or ovate-lanceolate, chiefly entire, rough on the margins, *the upper ones clasping by an auricled base.*—Dry woods.

8. **A. Novæ-An'gliæ,** L. Rays many, narrow, violet-purple; *heads large. Involucre of many slender equal scales, apparently in a single row, clammy.* Stem stout, 3-8 feet high, hairy, corymbed above. Leaves very numerous, lanceolate, entire, clasping by an auricled base, pubescent.—River-banks and borders of woods.

9. **A. puni'ceus,** L. *Rays long,* lilac-blue. *Scales of the involucre narrowly linear, loose.* Stem 3-6 feet high, stout, rough-hairy, *usually purple below.* Leaves oblong-lanceolate, clasping by an auricled base, sparingly serrate in the middle, rough above, smooth beneath, pointed.—Swamps ; usually clustered.

10. **A. longifo'lius,** Lam. Rays large, numerous, purplish-blue. *Scales of the involucre in several rows, linear, with awl-shaped spreading green tips.* Stem smooth. Leaves lanceolate or linear, taper-pointed, *shining above.* Heads solitary or few on the branchlets.—Moist thickets along streams.

* * * *None of the leaves heart-shaped ; those of the stem sessile, tapering at the base (except in No. 11).*

11. **A. multiflo'rus,** Ait. Rays white. Stem pale or hoary with minute pubescence, 1 foot high, bushy. *Leaves crowded, linear,* with rough margins ; the upper partly clasping. Heads crowded on the racemose branches. Scales of the involucre with spreading green tips.—Dry soil.

12. **A. Tradescan'ti,** L. Rays white or whitish. Scales of the involucre narrowly linear, in 3 or 4 rows. Heads small, very numerous, in 1-sided *close* racemes on the branches. Stem 2-4 feet high, bushy, *smooth.* Leaves linear-lanceolate, the larger ones with a *few remote teeth* in the middle.—Moist banks.

13. **A. miser,** L. (*A. diffusus,* Hook., in Macoun's Catalogue.) Rays pale blue or whitish. Involucre nearly as in No. 12. Stem *more or less hairy,* much branched. Heads small, in *loose* racemes

on the spreading branches. Leaves lanceolate, acute at each end, *sharply serrate in the middle.*—Low grounds.

14. **A. simplex**, Willd. Rays pale blue or whitish. Scales of the involucre linear-awl-shaped. Stem stout, smooth or nearly so, with numerous leafy branches. Heads medium-sized, somewhat corymbose. Leaves smooth, lanceolate, tapering at both ends, the lower serrate. —Moist and shady banks.

15. **A. tenuifo'lius**, L. Rays pale blue or whitish. Scales of the involucre linear-awl-shaped, with very slender points. Heads medium-sized, in panicled racemes. *Leaves long, narrowly lanceolate, tapering to a long slender point,* the lower usually serrate in the middle. Stem much branched, pubescent in lines. — Low thickets.

16. **A. nemora'lis**, Ait., is found in the eastern provinces and in Muskoka. Rays lilac-purple. Stem slender and leafy, the upper branches terminating in 1-flowered nearly naked peduncles. Leaves rigid, narrowly lanceolate, nearly entire, *with revolute margins.*

17. **A. ptarmicoi'des**, Torr. and Gr. Rays pure white. Stems clustered, generally a foot high, each bearing a flat corymb of small heads. Leaves linear-lanceolate, acute, rigid, entire, mostly 1-nerved, with rough margins.—Dry or gravelly hills. Our earliest Aster.

18. **A. acumina'tus**, Michx. Rays white or faintly purple. Stem about a foot high, somewhat hairy, zigzag, panicled-corymbose at the top. Leaves large, thin, oblong-lanceolate, pointed, coarsely toothed towards the apex, entire at the base.—Cool sandy woods ; mostly in eastern Ontario.

19. **ERIG'ERON**, L. FLEABANE.

1. **E. Canadense**, L. (HORSE-WEED. BUTTER-WEED.) Rays white, *but very inconspicuous,* shorter than their tubes. Heads very small, numerous, in panicled racemes. Stem 3-5 feet high, erect and wand-like, bristly-hairy. Leaves linear, mostly entire. Common in burnt woods and new clearings.

2. **E. bellidifo'lium**, Muhl. (ROBIN'S PLANTAIN.) Rays *bluish-purple*, numerous. Heads medium-sized, few, on slender corym-

bose peduncles. Stem hairy, *producing offsets from the base.*
Radical leaves spathulate or obovate, toothed above the middle;
stem-leaves oblong, *few*, sessile or partly clasping, entire.—
Thickets.

3. **E. Philadel'phicum**, L. (COMMON FLEABANE.) Rays *rose-
purple, very numerous and narrow.* Heads small, few, in corymbs.
Stem hairy, with *numerous* stem-leaves. Radical leaves spathulate and toothed; the upper ones clasping by a heart-shaped base, entire.—Moist grounds.

4. **E. strigo'sum**, Muhl. (DAISY FLEABANE.) Rays white, conspicuous, numerous. *Pappus plainly double.* Stem and leaves roughish with minute appressed hairs, or nearly smooth. Lower leaves spathulate and slender-petioled, *entire or nearly so*, the upper lanceolate, scattered.—Dry fields and meadows.

5. **E. an'nuum**, Pers. (LARGER DAISY FLEABANE.) Rays white, tinged with purple. Pappus double. Stem rough with spreading hairs. Leaves *coarsely toothed;* the lower ovate, tapering into a margined petiole; the upper ovate-lanceolate. Heads corymbed.—Fields and meadows.

20. **DIPLOPAP'PUS**, Cass. DOUBLE-BRISTLED ASTER.

D. umbella'tus, Torr. and Gr. Stem smooth, leafy to the top, tall, simple. Leaves lanceolate, long-pointed. Heads very numerous in flat compound corymbs.—Moist thickets.

21. **HELE'NIUM**, L. SNEEZE-WEED.

H. autumna'le, L. (SNEEZE-WEED.) Stem nearly smooth. Leaves lanceolate, toothed. Disk globular.—Low river- and lake-margins.

22. **LEUCAN'THEMUM**, Tourn. OX-EYE DAISY.

L. vulga're, Lam. (*Chrysanthemum Leucanthemum*, L., in Macoun's Catalogue.) (OX-EYE DAISY. WHITE-WEED.) Stem erect, naked above, bearing a single large head. Leaves pinnatifid or cut-toothed, the lowest spathulate, the others partly clasping.—Pastures and old fields.

23. **MARU'TA**, Cass. MAY-WEED.

M. Cot'ula, DC. (COMMON MAY-WEED.) Stem branching. Leaves thrice-pinnate, finely dissected.—Roadsides everywhere.

24. RUDBECKIA, L. CONE-FLOWER.

1. **R. lacinia'ta,** L. Rays linear, 1-2 inches long, drooping. *Disk greenish-yellow.* Stem tall, *smooth*, branching. Lowest leaves pinnate, of 5-7 lobed leaflets; upper ones 3-5-parted, or the uppermost undivided and generally ovate. Heads terminal, long-peduncled.—Swamps.

2. **R. hir'ta,** L. Rays bright yellow. *Disk purplish-brown.* Stem very *rough-hairy*, naked above, bearing single large heads. Leaves 3-ribbed, the lowest spathulate, narrowed into a petiole, the upper ones sessile.—Meadows.

25. HELIAN'THUS, L. SUN-FLOWER.

1. **H. strumo'sus,** L. Stem 3-4 feet high, smooth below. Leaves broadly lanceolate, rough above and whitish beneath, pointed, serrate with small appressed teeth, short-petioled. Rays about 10.—Moist copses and low grounds.

2. **H. divarica'tus,** L. Stem 1-4 feet high, *smooth*, simple or forking above. Leaves all opposite, widely spreading, *sessile, rounded or truncate at the base*, ovate-lanceolate, *3-nerved*, long-pointed, serrate, *rough on both sides*. Heads few, on short peduncles. Rays about 12.—Open thickets and dry plains.

3. **H. decapet'alus,** L. Stem 3-6 feet high, branching, smooth below, rough above. Leaves *thin, green on both sides, ovate, coarsely serrate*, pointed, abruptly contracted into short margined petioles. Rays usually 10.—Thickets and river-banks.

4. **H. gigante'us,** L. Stem tall, *hairy or rough*, branching above. Leaves lanceolate, pointed, serrate, very rough above, hairy below, narrowed and ciliate at the base. Heads somewhat corymbed, not large. Disk yellow; rays pale yellow, 15-20.—Low grounds, western and south-western Ontario.

5. **H. tubero'sus,** L., (JERUSALEM ARTICHOKE) has escaped from cultivation in some places. It is at once recognized by its tubers.

26. BIDENS, L. BUR-MARIGOLD.

1. **B. frondo'sa,** L. (COMMON BEGGAR-TICKS.) *Rays none. Achenes flat*, wedge-obovate, *ciliate on the margins with bristles pointing upwards*, 2-awned. Stem tall, branched. Leaves thin,

long-petioled, pinnately 3-5-divided, the leaflets ovate-lanceolate, pointed, serrate.

2. **B. conna'ta**, Muhl. (SWAMP BEGGAR-TICKS.) *Rays none.* Achenes flat, narrowly wedge-shaped, 2-4-awned, *ciliate with minute bristles pointing downwards.* Stem 1-2 feet high, smooth. Leaves lanceolate, pointed, serrate, tapering and connate at the base, the lowest often 3-parted and decurrent on the petiole. —In shallow water and low grounds.

3. **B. cer'nua, L.** (SMALLER BUR-MARIGOLD.) *Rays short, pale yellow.* Achenes flat, wedge-obovate, *4-awned, ciliate with bristles pointing downwards.* Stem nearly smooth, 5-10 inches high. Leaves all simple, lanceolate, unequally serrate, hardly connate. Heads nodding.—Wet places.

4. **B. chrysanthemoi'des**, Michx. (LARGER BUR-MARIGOLD.) *Rays an inch long, showy, golden yellow.* Achenes wedge-shaped, 2-4-awned, *bristly downwards.* Stem smooth, 6-30 inches high, erect or ascending. Leaves lanceolate, tapering at both ends, connate, regularly serrate.—Swamps and ditches.

5. **B. Beck'ii**, Torr. (WATER MARIGOLD.) *Aquatic.* Stems long and slender. Immersed leaves dissected into fine hair-like divisions; those out of water lanceolate, slightly connate, toothed. Rays showy, golden yellow, larger than the involucre. *Achenes linear, bearing 4-6 very long awns barbed towards the apex.*—Ponds and slow streams.

27. **HELIOP'SIS**, Pers. OX-EYE.

H. lævis, Pers. Stem smooth, slender, branching. Leaves ovate-lanceolate, acute, sharply serrate, on slender petioles. Heads showy; peduncles elongated.—Dry open thickets; London and westward.

28. **ACHILLE'A**, L. YARROW.

A. millefo'lium, L. (MILFOIL.) Stems simple. Leaves dissected into fine divisions. Corymb flat-topped. Rays only 4 or 5, short.—Fields and along fences; very common.

29. **POLYM'NIA**, L. LEAF-CUP.

P. Canadensis, L. A coarse clammy-hairy herb. Lower leaves opposite, petioled, pinnatifid; the upper alternate, angled.

or lobed. Heads small; rays pale yellow.—Shaded ravines, south-westward.

30. SIL'PIIIUM, L. ROSIN-PLANT.

1. **S. perfolia'tum**, L., (CUP-PLANT) is found in south-western Ontario. Stem stout, *square*, 4-8 feet high. Leaves ovate, coarsely toothed, the upper ones united by their bases.

2. **S. terebinthina'ceum**, L. (PRAIRIE DOCK.) Stem tall, *round*, naked above, smooth. Radical leaves sometimes 2 feet long, rough-hairy, coarsely serrate, on slender petioles. Heads small, loosely panicled.—Open woods and grassy banks, south-western Ontario.

31. CYNTHIA, Don. CYNTHIA.

C. Virgin'ica, Don. (*Krigia amplexicaulis*, Nutt., in Macoun's Catalogue.) Roots fibrous. Stem-leaves 1-2, oblong or lanceolate-spathulate, clasping, mostly entire, the radical ones on short winged petioles. Peduncles 2-5.—South-western Ontario.

32. LAMP'SANA, Tourn. NIPPLE-WORT.

L. commu'nis, L. Very slender and branching. Leaves angled or toothed. Heads small, loosely panicled.—Borders of springs; not common.

33. CICHO'RIUM, Tourn. SUCCORY. CICHORY.

C. In'tybus, L. Stem-leaves oblong or lanceolate, partly clasping; radical ones runcinate.—Roadsides and waste places.

34. LEON'TODON, L. FALL DANDELION.

L. autumna'le, L. (FALL DANDELION.) Leaves lanceolate, laciniate-toothed or pinnatifid. *Scape branched.*—Roadsides and waste places; not common.

35. HIERA'CIUM, Tourn. HAWKWEED.

1. **H. Canadensis**, Michx. (CANADA HAWKWEED.) *Heads large.* Stem simple, leafy, corymbed, 1-3 feet high. *Peduncles downy.* Leaves ovate-oblong, with a few coarse teeth, somewhat hairy, sessile, or the uppermost slightly clasping. Achenes tapering towards the base.—Dry banks and plains.

2. **H. scabrum**, Michx. (ROUGH H.) *Heads small.* Stem stout, 1-3 feet high, *rough-hairy*, corymbose. *Peduncles or involucre densely clothed with dark bristles.* Achenes not tapering.—Sandy woods and thickets.

3. **H. Gronovii**, L. (HAIRY H.) Heads small. Stem wand-like, leafy and very hairy below, *naked above*, forming a long and narrow panicle. *Achenes with a very taper summit.*—Dry soil, western Ontario.

4. **H. veno'sum**, L., (RATTLESNAKE-WEED) with a smooth naked scape (or bearing one leaf), and a loose corymb of very slender peduncles, is found in the Niagara region and southwestward.

36. **NAB'ALUS**, Cass. RATTLESNAKE-ROOT.

1. **N. albus**, Hook. (*Prenanthes alba*, L., in Macoun's Catalogue. (WHITE LETTUCE.) Heads 8-12-flowered. *Pappus deep cinnamon-coloured.* Stem 2-4 feet high, smooth and glaucous. corymbose-paniculate. Leaves triangular-halberd-shaped, or 3-5-lobed, the uppermost oblong and undivided.—Rich woods.

2. **N. altis'simus**, Hook. (*Prenanthes altissima*, L., in Macoun's Catalogue.) (TALL WHITE LETTUCE.) Heads 5-6-flowered. *Pappus pale straw-coloured.* Stem taller but more slender than in No. 1, *with a long leafy panicle at the summit.*—Rich woods.

3. **N. racemo'sus**, Hook. (*Prenanthes racemosa*, Michx., in Macoun's Catalogue.) Heads about 12-flowered. Involucre and peduncles *hairy.* Stem wand-like, smooth. Leaves oval or oblong-lanceolate, slightly toothed. Heads crowded in a long and narrow interruptedly spiked panicle. Pappus straw-colour; flowers flesh-colour.—Shore of Lake Huron and south-westward.

37. **TARAX'ACUM**, Haller. DANDELION.

T. Dens-leo'nis, Desf. (COMMON DANDELION.) Outer involucre reflexed. Leaves runcinate.—Fields everywhere.

38. **LACTU'CA**, Tourn. LETTUCE.

1. **L. Canadensis**, L. (WILD LETTUCE.) Heads numerous, in a long and narrow naked panicle. Stem stout, smooth, hollow, 4-9 feet high. Leaves mostly runcinate, partly clasping, pale beneath; the upper entire. Achenes longer than their beaks.—Borders of fields and thickets.

2. **L. integrifo'lia,** L. Stem 3-6 feet high; leaves all undivided, entire or slightly toothed. Flowers pale yellow, cream-colour, or purple.—Dry soil.

3. **L. hirsu'ta,** Muhl. Leaves runcinate, the midrib beneath often sparingly bristly-hairy. Flowers yellowish-purple, rarely white.—Dry soil.

39. MULGE'DIUM, Cass. FALSE OR BLUE LETTUCE.

M. leucophæ'um, DC. (*Lactuca leucophæa*, Gray, in Macoun's Catalogue.) Stem tall and very leafy. Heads in a dense compound panicle.—Borders of damp woods and along fences.

40. SON'CHUS, L. SOW-THISTLE.

1. **S. olera'ceus,** L. (COMMON SOW-THISTLE.) Stem-leaves runcinate, slightly toothed with soft spiny teeth, clasping; *the auricles acute*.—Manured soil about dwellings.

2. **S. asper,** Vill. (SPINY-LEAVED S.) Leaves *hardly lobed*, fringed with soft spines, clasping; *the auricles rounded*. Achenes *margined*.—Same localities as No. 1.

3. **S. arven'sis,** L., (FIELD S.) with bright yellow flowers and bristly involucres and peduncles, is found eastward.

41. TRAGOPO'GON, L. GOAT'S BEARD. SALSIFY.

T. praten'sis, L. (YELLOW GOAT'S BEARD.) Spreading westward along the railway lines.

ORDER L. LOBELIA'CEÆ. (LOBELIA FAMILY.)

Herbs, with milky acrid juice, alternate leaves, and loosely racemed flowers. Corolla irregular, 5-lobed, *the tube split down one side*. Stamens 5, syngenesious, and commonly also monadelphous, *free from the corolla*. Calyx-tube adherent to the many-seeded ovary. Style 1. The only genus is

LOBE'LIA, L. LOBELIA.

1. **L. cardina'lis,** L. (CARDINAL-FLOWER.) Corolla *large, deep red*. Stem simple, 2-3 feet high, smooth. Leaves oblong-lanceolate, slightly toothed. Bracts of the flowers leaf-like.—Low grounds.

2. **L. syphilit'ica,** L. (GREAT LOBELIA.) Corolla *rather large, light blue.* Stem hairy, simple, 1-2 feet high. Leaves thin, acute at both ends, serrate. Calyx-lobes half as long as the corolla, the tube hemispherical. Flowers in a dense spike or raceme.—Low grounds.

3. **L. infla'ta,** L. (INDIAN TOBACCO.) Flowers *small,* ⅛ *of an inch long, pale blue.* Stem leafy, *branching,* 8-18 inches high, pubescent. Leaves ovate or oblong, toothed. *Pods inflated. Racemes leafy.*—Dry fields.

4. **L. spica'ta,** Lam. Flowers small, ⅓ *of an inch long, pale blue.* Stem slender, erect, *simple,* 1-3 feet high, minutely pubescent below. Leaves barely toothed, the lower spathulate or obovate, the upper reduced to linear bracts.—*Racemes long and naked.*—Sandy soil.

5. **L. Kal'mii,** L. Flowers small, ⅛ of an inch long, *light blue.* Stem low, 4-18 inches high, *very slender.* Pedicels filiform, as long as the flowers, with 2 minute bractlets above the middle. Leaves mostly linear, the radical ones spathulate and the uppermost reduced to bristly bracts.—Wet rocks and banks, chiefly northward.

6. **L. Dortman'na,** L., (WATER LOBELIA) with small leaves all tufted at the root, and a scape 5 or 6 inches long with a few small light-blue pedicelled flowers at the summit, occurs in the shallow borders of ponds in Muskoka.

ORDER LI. **CAMPANULA'CEÆ.** (CAMPANULA F.)

Herbs with milky juice, *differing from the preceding Order chiefly in having a regular 5-lobed corolla (bell-shaped or wheel-shaped), separate stamens (5), and 2 or more (with us, 3) stigmas.*

Synopsis of the Genera.

1. Campan'ula. Calyx 5-cleft. Corolla generally bell-shaped, 5-lobed. Pod short.
2. Specula'ria. Calyx 5-cleft. Corolla nearly wheel-shaped, 5-lobed. Pod prismatic or oblong.

1. **CAMPAN'ULA,** Tourn. BELL-FLOWER.

1. **C. rotundifo'lia,** L. (HAREBELL.) Flowers blue, loosely panicled, on long slender peduncles, nodding. Stem slender,

branching, several-flowered. Root-leaves round-heart-shaped; stem-leaves linear. Calyx-lobes awl-shaped.—Shaded cliffs.

2. **C. aparinoi'des**, Pursh. (MARSH BELL-FLOWER.) Flowers *white* or nearly so, about ⅓ of an inch long. *Stem very slender and weak, few-flowered, angled, roughened backwards.* Leaves linear-lanceolate. Calyx-lobes triangular.—Wet places in high grass. The plant has the habit of a Galium.

3. **C. America'na**, L. (TALL BELL-FLOWER.) Flowers light blue, about an inch across, *crowded in a leafy spike.* Corolla deeply 5-lobed. *Style long and curved.* Stem 3-6 feet high, simple. Leaves ovate or ovate-lanceolate, taper-pointed, serrate. —Moist rich soil.

2. SPECULA'RIA, Heister. VENUS'S LOOKING-GLASS.

S. perfolia'ta, A. DC. Flowers purplish-blue, only the latter or upper ones expanding. Stem hairy, 3-20 inches high. Leaves roundish or ovate, clasping. Flowers solitary or 2 or 3 together in the axils.—Sterile open ground, chiefly south-westward.

ORDER LII. **ERICA'CEÆ.** (HEATH FAMILY.)

Chiefly shrubs, *distinguished by the anthers opening, as a rule, by a pore at the top of each cell.* Stamens (as in the two preceding Orders) free from the corolla, as many or twice as many as its lobes. Leaves simple and usually alternate. Corolla in some cases polypetalous.

Synopsis of the Genera.

SUBORDER I. **VACCINIEÆ.** (WHORTLEBERRY FAMILY.)

Calyx-tube adherent to the ovary. Fruit a berry crowned with the calyx-teeth.

1. **Gaylussa'cia.** Stamens 10, the anthers opening by a pore at the apex. Corolla tubular, ovoid, the border 5-cleft. *Berry 10-celled, 10-seeded.* Flowers white with a red tinge. Leaves covered with resinous dots. Branching shrubs.

2. **Vaccin'ium.** Stamens 8 or 10, the anthers prolonged upwards into tubes with a pore at each apex. Corolla deeply 4-parted and revolute, or cylindrical with the limb 5-toothed. Berry 4-celled, or more or less completely 10-celled. Flowers white or reddish, solitary or in short racemes. Shrubs.

3. **Chiog'enes.** Stamens 8, *each anther 2-pointed at the apex.* Corolla bell-shaped, *deeply 4-cleft.* Limb of the calyx 4-parted. Flowers very small, nodding from the axils, with 2 bractlets under the calyx. *Berry white,* 4-celled. A trailing slender evergreen.

SUBORDER II. **ERICINEÆ.** (HEATH FAMILY PROPER.)

Calyx free from the ovary. Shrubs or small trees. Corolla gamopetalous, except in No. 10.

4. **Arctostaph'ylos.** Corolla urn-shaped, the limb 5-toothed, revolute. Stamens 10, the anthers each with 2 reflexed awns on the back. Fruit a berry-like drupe, 5-10-seeded. A trailing thick-leaved evergreen with nearly white flowers.

5. **Epigæ'a.** Corolla salver-shaped, hairy inside, rose-coloured. Stamens 10; filaments slender ; anthers awnless, *opening lengthwise.* Calyx of 5 pointed and scale-like nearly distinct sepals. A trailing evergreen, bristly with rusty hairs.

6. **Gaulthe'ria.** Corolla ovoid, or slightly urn-shaped, 5-toothed, nearly white. Stamens 10, the anthers 2-awned. Calyx 5-cleft, enclosing the pod and *becoming fleshy and berry-like in fruit.* Stems low and slender, leafy at the summit.

7. **Cassan'dra.** Corolla cylindrical, 5-toothed. Stamens 10, the anther-cells tapering into beaks with a pore at the apex, awnless. Calyx of 5 overlapping sepals, and 2 similar bractlets. Pod with a double pericarp, the outer of 5 valves, the inner cartilaginous and of 10 valves. A low shrub with rather scurfy leaves, and white flowers.

8. **Androm'eda.** Corolla globular-urn-shaped, 5-toothed. Calyx of 5 nearly distinct valvate sepals, without bractlets. Stamens 10; the filaments bearded ; the anther-cells each with a slender awn. A low shrub, with white flowers in a terminal umbel.

9. **Kal'mia.** Corolla broadly bell-shaped, *with 10 pouches receiving as many anthers.* Shrubs with showy rose-purple flowers.

10. **Le'dum.** Calyx 5-toothed, very small. Corolla of 5 obovate and spreading *distinct petals.* Stamens 5-10. Leaves evergreen, *with revolute margins, covered beneath with rusty wool.*

SUBORDER III. **PYROLEÆ.** (PYROLA FAMILY.)

Calyx free from the ovary. Corolla polypetalous. More or less herbaceous evergreens.

11. **Py'rola.** Calyx 5-parted. Petals 5, concave. Stamens 10. Stigma 5-lobed. Leaves evergreen, *clustered at the base of an upright scaly-bracted scape which bears a simple raceme of nodding flowers.*

12. **Mone'ses.** Petals 5, orbicular, spreading. Stamens 10. Stigma large, peltate, with 5 narrow radiating lobes. Plant having the aspect of a Pyrola, *but the scape bearing a single terminal flower.*

13. **Chimaph'ila.** Petals 5, concave, orbicular, spreading. Stamens 10. Stigma broad and round, the border 5-crenate. Low plants with running underground shoots, and thick, shining, sharply serrate, somewhat whorled leaves. Flowers corymbed or umbelled on a terminal peduncle.

SUBORDER IV. **MONOTROPEÆ.** (INDIAN-PIPE FAMILY.)

14. **Monot'ropa.** A smooth *perfectly white* plant, parasitic on roots, bearing scales instead of leaves, and a single flower at the summit of the stem.
15. **Pteros'pora.** A purplish-brown clammy-pubescent plant, *parasitic* on the roots of pines. Stem simple. Flowers numerous, nodding, white, forming a raceme.
16. **Hypop'itys.** A tawny or reddish *parasitic* plant, with several flowers in a scaly raceme, the terminal one generally with 5 petals and 10 stamens, and the others with 4 petals and 8 stamens.

1. GAYLUSSA'CIA, H.B.K. HUCKLEBERRY.

G. **resino'sa**, Torr. and Gr. (BLACK HUCKLEBERRY.) Fruit *black, without a bloom*. Racemes short, 1-sided, in clusters. Leaves oval or oblong. Branching shrub, 1-3 feet high.—Low grounds.

2. VACCIN'IUM, L. CRANBERRY. BLUEBERRY.

1. **V. Oxycoc'cus,** L. (*Oxycoccus vulgaris*, Pursh, in Macoun's Catalogue.) (SMALL CRANBERRY.) A creeping or trailing very slender shrubby plant, with ovate acute evergreen leaves only ¼ of an inch long, the margins revolute. Corolla rose-coloured, 4-parted, the lobes reflexed. Anthers 8. *Stem 4-9 inches long.* Berry only about ⅓ of an inch across, often speckled with white. —Bogs.

2. **V. macrocar'pon,** Ait. (*Oxycoccus macrocarpus*, Pursh, in Macoun's Catalogue.) (LARGE or AMERICAN CRANBERRY.) Different from No. 1 in having prolonged stems (1-3 feet long) and the flowering branches lateral. The leaves also are nearly twice as large, and the *berry is fully ½ an inch broad.*—Bogs.

3. **V. Vitis-Idæa,** L. A low plant with erect branches from tufted creeping stems. Leaves evergreen, *obovate*, with revolute margins, shining above, dotted with blackish bristly points beneath. Corolla *bell-shaped*, 4-lobed. Anthers 8-10. Flowers in a short bracted raceme.—Northward and eastward.

4. **V. Pennsylva'nicum,** Lam. (DWARF BLUEBERRY.) Stem 6-15 inches high, the branches green, angled, and warty. Corolla

cylindrical bell-shaped, 5-toothed. Anthers 10. Flowers in short racemes. Leaves lanceolate or oblong, serrulate with bristly-pointed teeth, smooth and shining on both sides. Berry blue or black, with a bloom.—Dry plains and woods.

5. **V. Canadense**, Kalm. (CANADIAN BLUEBERRY.) Stem 1-2 feet high. Leaves oblong-lanceolate or elliptical, *entire*, *downy both sides*, as are also the branchlets.—A very common Canadian species.

6. **V. corymbo'sum**, L., (SWAMP-BLUEBERRY) is a tall shrub (3-10 feet) growing in swamps and low grounds, with leaves varying from ovate to elliptical-lanceolate, and flowers and berries very much the same as those in No. 4, but the berries ripen later.

3. CHIOG'ENES, Salisb. CREEPING SNOWBERRY.

C. hispid'ula, Torr. and Gr. Leaves very small, ovate and pointed, on short petioles, the margins revolute. The lower surface of the leaves and the branches clothed with rusty bristles. Berries bright white.—Bogs and cool woods.

4. ARCTOSTAPH'YLOS, Adans. BEARBERRY.

A. Uva-ursi, Spreng. Flowers in terminal racemes. Leaves alternate, obovate or spathulate, entire, smooth. Berry red. Bare hillsides.

5. EPIGÆ'A, L. GROUND LAUREL. TRAILING ARBUTUS.

E. re'pens, L. (MAYFLOWER.) Flowers in small axillary clusters from scaly bracts. Leaves evergreen, rounded and heart-shaped, alternate, on slender petioles. Flowers very fragrant.—Dry woods in early spring.

6. GAULTHE'RIA, Kalm. AROMATIC WINTERGREEN.

G. procum'bens, L. (TEABERRY. WINTERGREEN.) Flowers mostly single in the axils, nodding. Leaves obovate or oval, obscurely serrate, evergreen. Berry bright red, edible.—Cool woods, chiefly in the shade of evergreens.

7. CASSAN'DRA, Don. LEATHER-LEAF.

C. calycula'ta, Don. Flowers in 1-sided leafy racemes. Leaves oblong, obtuse, flat.—Bogs.

ERICACEÆ.

8. ANDROM'EDA, L. ANDROMEDA.

A. polifo'lia, L. Stem smooth and glaucous, 6-18 inches high. Leaves oblong-linear, with strongly revolute margins, white beneath.—Bogs.

9. KAL'MIA, L. AMERICAN LAUREL.

K. glau'ca, Ait. (PALE LAUREL.) A straggling shrub about a foot high, with few-flowered terminal corymbs. Branchlets 2-edged. Leaves opposite, oblong, the margins revolute. Flowers ½ an inch across.—Bogs.

10. LE'DUM, L. LABRADOR TEA.

L. latifo'lium, Ait. Flowers white, in terminal umbel-like clusters. Leaves elliptical or oblong. Stamens 5, or occasionally 6 or 7.—Bogs.

11. PY'ROLA, Tourn. WINTERGREEN. SHIN-LEAF.

1. **P. rotundifo'lia,** L. *Leaves orbicular, thick, shining, usually shorter than the petiole.* Calyx-lobes lanceolate. Flowers white, or in var. **incarna'ta** rose-purple.—Moist woods.

2. **P. ellip'tica,** Nutt. (SHIN-LEAF.) *Leaves elliptical, thin, dull, usually longer than the margined petiole.* Flowers greenish-white.—Rich woods.

3. **P. secunda,** L. Easily recognized by the *flowers of the dense raceme being all turned to one side.* Leaves ovate. Style long, protruding.—Rich woods. Var. **pumila** has *orbicular* leaves, and is 3-8-flowered.—Peat-bogs and swamps.

4. **P. chloran'tha,** Swartz, has small roundish dull leaves, converging greenish-white petals, and *the anther-cells contracted below the pore into a distinct neck or horn.*—Open woods.

12. MONE'SES, Salisb. ONE-FLOWERED PYROLA.

M. uniflo'ra, Gr. Leaves thin, rounded, veiny, and serrate. Scape 2-4 inches high, bearing a single white or rose-coloured flower.—Deep woods.

13. CHIMAPH'ILA, Pursh. PIPSISSEWA.

1. **C. umbella'ta,** Nutt. (PRINCE'S PINE.) Leaves wedge-lanceolate, *acute at the base.* Peduncles 4-7-flowered. Corolla rose- or flesh-coloured.—Dry woods.

2. **C. macula'ta,** Pursh. (SPOTTED WINTERGREEN.) Leaves ovate-lanceolate, *obtuse at the base,* the upper surface *variegated with white.*—Dry woods.

14. MONOT'ROPA, L. INDIAN PIPE. PINE-SAP.

M. uniflo'ra, L. (INDIAN PIPE. CORPSE-PLANT.) Smooth, waxy-white, turning black in drying.—Dark rich woods.

15. PTEROS'PORA. Nutt. PINE-DROPS.

P. Andromede'a, Nutt. Calyx 5-parted. Corolla ovate, urn-shaped, 5-toothed, persistent. Stamens 10. Stigma 5-lobed. Pod 5-lobed, 5-celled.—Usually under pines in dry woods.

16. HYPOP'ITYS, Scop. PINE-SAP.

H. lanugino'sa, Nutt. Somewhat pubescent. Sepals bract-like. Stigma ciliate. Style longer than the ovary, hollow. Pod globular or oval.—Oak and pine woods.

ORDER LIII. **AQUIFOLIA'CEÆ.** (HOLLY FAMILY.)

Shrubs or small trees, with small axillary polygamous or diœcious flowers, the parts mostly in fours or sixes. Calyx very minute, free from the ovary. Stamens alternate with the petals, attached to their base, the corolla being almost polypetalous. Anthers opening lengthwise. Stigma nearly sessile. Fruit a berry-like 4-8-seeded drupe.

1. ILEX, L. HOLLY.

I. verticilla'ta, Gr. (BLACK ALDER. WINTERBERRY.) A shrub with the greenish flowers in sessile clusters, or the fertile ones solitary. Parts of the flowers mostly in sixes. Fruit bright red. Leaves alternate, obovate, oval, or wedge-lanceolate, pointed, veiny, serrate.—Swamps and low grounds.

2. NEMOPAN'THES, Raf. MOUNTAIN HOLLY.

N. Canadensis, DC. A branching shrub, with grey bark, and alternate oblong nearly entire smooth leaves on slender petioles. Flowers on long slender axillary peduncles, mostly solitary. Petals 4-5, oblong-linear, *distinct*. Fruit light red.— Moist woods.

Order LIV. **PLANTAGINA'CEÆ.** (Plantain Family.)

Herbs, with the leaves all radical, and the flowers in a close spike at the summit of a naked scape. Calyx of 4 sepals, persistent. Corolla 4-lobed, thin and membranaceous, spreading. Stamens 4, usually with long filaments, inserted on the corolla. Pod 2-celled, the top coming off like a lid. Leaves ribbed. The principal genus is

PLANTA'GO, L. Plantain. Rib-Grass.

1. **P. major**, L. (Common P.) *Spike long and slender. Leaves 5-7-ribbed*, ovate or slightly heart-shaped, *with channelled petioles. Pod 7-16-seeded.*—Moist ground about dwellings.

2. **P. Kamtscha'tica**, Hook. (*P. Rugelii*, Decaisne, in Macoun's Catalogue.) Resembling small forms of No. 1, but *pod 4-seeded.*

3. **P. lanceola'ta**, L. (Rib-Grass. English Plantain.) *Spike thick and dense, short.* Leaves 3-5-ribbed, lanceolate or lanceolate-oblong. Scape grooved, long and slender.—Dry fields and banks.

4. **P. corda'ta**, Lam. Tall and glabrous. *Bracts round-ovate, fleshy.*—Pod 2-4-seeded.—South-western Ontario.

5. **P. marit'ima**, L., var. **juncoi'des**, Gr., with very narrow and slender spike, and linear fleshy leaves, is found on the sea-coast and Lower St. Lawrence.

Order LV. **PRIMULA'CEÆ.** (Primrose Family.)

Herbs with regular perfect flowers, well marked by having *a stamen before each petal or lobe of the corolla* and inserted on the tube. Ovary 1-celled, the placenta rising from the base. Style 1; stigma 1.

Synopsis of the Genera.

1. **Prim'ula.** Leaves all in a cluster at the root. Flowers in an umbel at the summit of a simple scape. Corolla salver-shaped. Stamens 5, included.

2. **Trienta'lis.** Leaves in a whorl at the summit of a slender erect stem. Calyx usually 7-parted, the lobes pointed. Corolla usually 7-parted, spreading, without a tube. Filaments united in a ring below. Flowers usually only one, white and star-shaped.

3. **Lysimach'ia.** Leafy-stemmed. Flowers yellow, axillary or in a terminal raceme. Calyx usually 5-parted. Corolla wheel-shaped, mostly 5-parted, and sometimes polypetalous.

4. **Anagal'lis.** Low and spreading. Leaves opposite or whorled, entire. Flowers variously coloured, solitary in the axils. Calyx 5-parted. Corolla wheel-shaped, 5-parted. Filaments bearded.

5. **Sam'olus.** Smooth and spreading, 6-10 inches high. Corolla bell-shaped, 5-parted, with 5 sterile filaments in the sinuses. *Calyx partially adherent to the ovary.* Flowers very small, white, racemed. *Leaves alternate.*

1. PRIM'ULA, L. PRIMROSE. COWSLIP.

1. **P. farino'sa, L.** (BIRD'S-EYE P.) Lower surface of the leaves covered with a white mealiness. Corolla lilac with a yellow centre.—Shores of Lake Huron, and northward.

2. **P. Mistassin'ica,** Michx. Leaves not mealy. Corolla flesh-coloured, the lobes obcordate.—Shores of the Upper Lakes, and northward.

2. TRIENTA'LIS, L. CHICKWEED-WINTERGREEN.

T. America'na, Pursh. (STAR-FLOWER.) Leaves thin and veiny, lanceolate, tapering towards both ends. Petals pointed. —Moist woods.

3. LYSIMACH'IA, Tourn. LOOSESTRIFE.

1. **L. thyrsiflo'ra, L.** (TUFTED LOOSESTRIFE.) Flowers in spike-like clusters from the axils of a few of the upper leaves. Petals lance-linear, *purplish-dotted*, as many minute teeth between them. Leaves scale-like below, the upper lanceolate, opposite, sessile, dark-dotted.—Wet swamps.

2. **L. stricta,** Ait. Flowers on slender pedicels *in a long terminal raceme.* Petals lance-oblong, streaked with dark lines. *Leaves opposite,* lanceolate, acute at each end, sessile, dark-dotted. —Low grounds.

3. **L. quadrifo'lia, L.** *Flowers on long slender peduncles from the axils of the upper leaves.* Petals streaked. *Leaves in whorls* of 4 or 5, ovate-lanceolate, dark-dotted.—Sandy soil.

4. **L. cilia'ta, L.** (*Steironema ciliatum,* Raf., in Macoun's Catalogue.) Flowers nodding on slender peduncles from the upper axils. *Petals not streaked or dotted.* Leaves opposite, *not*

...*sorted*, ovate-lanceolate, pointed, cordate at the base, *on long fringed petioles.*—Low grounds.

5. **L. longifo'lia**, Walt. (*Steironema longifolium*, Gray, in Macoun's Catalogue.) Petals not streaked or dotted. *Stem-leaves sessile, narrowly linear, 2-4 inches long*, the margins sometimes revolute. Stem 4-angled.—Moist soil, western Ontario.

4. **ANAGAL'LIS**, Tourn. PIMPERNEL.

A. arven'sis, L. (COMMON PIMPERNEL.) Petals obovate, *fringed with minute teeth*, mostly bluish or purplish. Flowers closing at the approach of rain. Leaves ovate, sessile.—Sandy fields and garden soil.

5. **SAM'OLUS**, L. WATER-PIMPERNEL. BROOK-WEED.

S. Valeran'di, L., var. **America'nus**, Gray. Stem slender, diffusely branched. The slender pedicels each with a bractlet at the middle.—Wet places.

ORDER LVI. **LENTIBULA'CEÆ.** (BLADDERWORT F.)

Small aquatic or marsh herbs, with a 2-lipped calyx and a personate corolla with a spur or sac underneath. Stamens 2. Ovary as in Primulaceæ. Chiefly represented by the two following genera:—

1. **UTRICULA'RIA**, L. BLADDERWORT.

1. **U. vulga'ris**, L. (GREATER BLADDERWORT.) Immersed leaves crowded, finely dissected into capillary divisions, furnished with small air-bladders. Flowers yellow, several in a raceme on a naked scape. Corolla closed; the spur conical and shorter than the lower lip.—Ponds and slow waters.

2. **U. interme'dia**, Hayne. Immersed leaves 4 or 5 times forked, the divisions linear-awl-shaped, minutely bristle-toothed on the margin, *not bladder-bearing*, the bladders being on leafless branches. Stem 3-6 inches long. Scape very slender, 3-6 inches long, bearing few yellow flowers. Upper lip of the corolla much longer than the palate; *the spur closely pressed to the broad lower lip.*—Shallow waters.

3. **U. cornu'ta**, Michx., with an awl-shaped spur turned downward and outward, and the lower lip of the corolla helmet-shaped,

is not uncommon in the northern parts of Ontario. Flowers yellow. Leaves awl-shaped.

2. PINGUIC'ULA, L. BUTTERWORT.

P. vulga'ris, L. A small and stemless perennial growing on damp rocks. Scapes 1-flowered. Leaves entire, ovate or elliptical, soft-fleshy, clustered at the root. Upper lip of the calyx 3-cleft, the lower 2-cleft. Corolla violet, the lips very unequal, the palate open, and hairy or spotted.—Shore of Lake Huron.

ORDER LVII. OROBANCHA'CEÆ. (BROOMRAPE F.)

Parasitic herbs, destitute of green foliage. Corolla more or less 2-lipped. Stamens didynamous. Ovary 1-celled with 2 or 4 parietal placentæ, many-seeded.

1. EPIPHE'GUS, Nutt. BEECH-DROPS.

E. Virginia'na, Bart. A yellowish-brown branching plant, parasitic on the roots of beech-trees. Flowers racemose or spiked; the upper sterile, with long corolla; the lower fertile, with short corolla.

2. CONOPH'OLIS, Wallroth. SQUAW-ROOT.

C. America'na, Wallroth. A chestnut-coloured or yellow plant found in clusters in oak woods in early summer, 3-6 inches high and rather less than an inch in thickness. The stem covered with fleshy scales so as to resemble a cone. Flowers under the upper scales; stamens projecting.

3. APHYL'LON, Mitchell. NAKED BROOM-RAPE.

A. uniflo'rum, Torr. and Gr. Plant yellowish-brown. Flower solitary at the top of a naked scape. Stem subterranean or nearly so, short and scaly. Scapes 3-5 inches high. Calyx 5-cleft, the divisions lance-awl-shaped. Corolla with a long curved tube and 5-lobed border, and 2 yellow-bearded folds in the throat. Stigma 2-lipped.—Woods, in early summer.

ORDER LVIII. SCROPHULARIA'CEÆ. (FIGWORT F.)

Herbs, distinguished by a 2-lipped or more or less irregular corolla, stamens usually 4 and didynamous, or only 2, (or in Verbascum 5) and a 2-celled and usually many-seeded ovary. Style 1; stigma entire or 2-lobed.

SCROPHULARIACEÆ. 95

Synopsis of the Genera.

* *Corolla wheel-shaped, and only slightly irregular.*

1. **Verbas'cum.** *Stamens (with anthers) 5.* Flowers in a long terminal spike. Corolla 5-parted, nearly regular. Filaments (or some of them) woolly.

2. **Veron'ica.** *Stamens only 2;* filaments long and slender. Corolla mostly 4-parted, nearly or quite regular. Pod flattish. Flowers solitary in the axils, or forming a terminal raceme or spike.

** *Corolla 2-lipped, or tubular and irregular.*

+ *Upper lip of the corolla embracing the lower in the bud, except occasionally in Mimulus.*

3. **Lina'ria.** *Corolla personate* (Fig. 181, Part I.), *with a long spur beneath.* Stamens 4. Flowers yellow, in a crowded raceme.

4. **Scrophula'ria.** Corolla tubular, somewhat inflated, 5-lobed; *the 4 upper lobes erect,* the lower one spreading. Stamens with anthers 4, the rudiment of a fifth in the form of a scale on the upper lip of the corolla. Flowers small and dingy, forming a narrow terminal panicle. Stem 4-sided.

5. **Chelo'ne.** Corolla inflated-tubular (Fig. 180, Part I.). Stamens 4, with woolly filaments and anthers, and a fifth filament without an anther. Flowers white, in a close terminal spike.

6. **Pentste'mon.** Corolla 2-lipped, gradually widening upwards. Stamens 4, with a fifth sterile filament, *the latter yellow-bearded.* Flowers white or purplish, in a loose panicle.

7. **Mim'ulus.** *Calyx 5-angled* and 5-toothed. Upper lip of the corolla erect or reflexed-spreading, the lower spreading, 3-lobed. Stamens 4, alike; *no rudiment of a fifth.* Stigma 2-lipped. Flowers blue or yellow, solitary on axillary peduncles.

8. **Grati'ola.** Corolla tubular and 2-lipped. *Stamens with anthers only 2,* included. Flowers with a yellowish tube, on axillary peduncles, solitary. Style dilated at the apex.

9. **Ilysan'thes.** Corolla tubular and 2-lipped. *Stamens with anthers only 2,* included; *also a pair of filaments which are two-lobed, but without anthers.* Flowers purplish, axillary. Style 2-lipped at the apex.

+ + *Lower lip of the corolla embracing the upper in the bud.*

10. **Gerar'dia.** Corolla funnel-form, swelling above, *the 5 spreading lobes more or less unequal.* Stamens 4, strongly didynamous, hairy. Style long, enlarged at the apex. Flowers purple or yellow, solitary on axillary peduncles, or sometimes forming a raceme.

11. **Castille'ia.** Corolla tubular and 2-lipped, *its tube included in the tubular and flattened calyx;* the upper lip long and narrow and flattened laterally, the lower short and 3-lobed. Stamens 4, didynamous. *Floral leaves scarlet* (rarely yellow) in our species. Corolla pale yellow.

12. **Euphra'sia.** *Calyx 4-cleft.* Upper lip of the corolla erect, the lower spreading. Stamens 4, under the upper lip. Very small herbs, with whitish or bluish spiked flowers. (Chiefly on the sea-coast, and north of Lake Superior.)

13. **Rhinan'thus.** *Calyx flat, greatly inflated in fruit,* 4-toothed. Upper lip of the corolla arched, flat, with a minute tooth on each side below the apex. Stamens 4. Flowers yellow, solitary in the axils, nearly sessile, the whole forming a crowded 1-sided spike. (Chiefly on the sea-coast, and north of Lake Superior.)

14. **Pedicula'ris.** Calyx split in front, not inflated in fruit. Corolla 2-lipped, the upper lip arched or hooded, incurved, flat, 2-toothed under the apex. Stamens 4. *Pod flat, somewhat sword-shaped.*

15. **Melampy'rum.** *Calyx 4-cleft, the lobes sharp-pointed.* Corolla greenish-yellow; upper lip arched, compressed, the lower 3-lobed at the apex. Stamens 4; anthers hairy. Pod 1-4-seeded, flat, oblique. Upper leaves larger than the lower ones and fringed with bristly teeth at the base.

1. VERBAS'CUM, L. MULLEIN.

1. **V. Thap'sus, L.** (COMMON MULLEIN.) A tall and very woolly herb, with the simple stem winged by the decurrent bases of the leaves. Flowers yellow, forming a dense spike.—Fields and roadsides everywhere.

2. **V. Blatta'ria, L.** (MOTH M.) Stem slender, *nearly smooth*. Lower leaves petioled, doubly serrate; the upper partly clasping. Flowers whitish with a purple tinge, *in a loose raceme. Filaments all violet-bearded.*—Roadsides; not common northward.

2. VERON'ICA, L. SPEEDWELL.

1. **V. America'na,** Schweinitz. (AMERICAN BROOKLIME.) Flowers pale blue, in *opposite* axillary racemes. *Leaves mostly petioled,* thickish, serrate. *Pod swollen.*—A common plant in brooks and ditches.

2. **V. Anagal'lis, L.,** (WATER SPEEDWELL) is much like No. 1, but *the leaves are sessile,* with a heart-shaped base.

3. **V. scutella'ta, L.** (MARSH S.) Flowers pale blue, in racemes chiefly from *alternate* axils. *Leaves sessile,* linear, opposite, hardly toothed. Racemes 1 or 2, *slender and zigzag.* Flowers few. Pods very flat, notched at both ends.—Bogs.

4. **V. officina'lis, L.** (COMMON S.) Flowers light blue. Stem prostrate, rooting at the base, *pubescent.* Leaves short-petioled,

obovate-elliptical, serrate. *Racemes dense*, chiefly from alternate axils. Pod obovate-triangular, notched.—Hillsides and open woods.

5. **V. serpyllifo'lia**, L. (THYME-LEAVED S.) Flowers whitish or pale blue, in a loose *terminal* raceme. Stem nearly smooth, branched at the creeping base. Leaves obscurely crenate, *the lowest petioled*. Pod flat, notched.—Roadsides and fields. Plant only 2 or 3 inches high.

6. **V peregri'na**, L. (NECKWEED.) Flowers whitish, *solitary in the axils of the upper leaves*. Whitish corolla, shorter than the calyx. Stem 4-9 inches high, *nearly smooth*. Pod orbicular, slightly notched.—Waste places and cultivated grounds.

7. **V. arven'sis**, L. (CORN SPEEDWELL.) Flowers (blue) as in No. 6, but *the stem is hairy*, and the *pod inversely heart-shaped*.—Cultivated soil.

3. LINA'RIA, Tourn. TOAD-FLAX.

L. vulga'ris, Mill. (TOAD-FLAX. BUTTER-AND-EGGS.) Leaves crowded, linear, pale green. Corolla pale yellow, with a deeper yellow or orange-coloured palate.—Roadsides.

4. SCROPHULA'RIA, Tourn. FIGWORT.

S. nodo'sa, L. Stem smooth, 3-4 feet high. Leaves ovate or oblong, the upper lanceolate, serrate.—Damp thickets.

5. CHELO'NE, Tourn. TURTLE-HEAD.

C. glabra, L. Stem smooth, erect and branching. Leaves short-petioled, lance-oblong, serrate, opposite. Bracts of the flowers concave.—Wet places.

6. PENTSTE'MON, Mitchell. BEARD-TONGUE.

P. pubes'cens. Stem 1-3 feet high, pubescent; the panicle more or less clammy. Throat of the corolla almost closed. Stem-leaves lanceolate, clasping.—Dry soil.

7. MIM'ULUS, L. MONKEY-FLOWER.

1. **M. rin'gens**, L. Stem square, 1-2 feet high. *Corolla blue*, an inch long. (A white-flowered variety is sometimes met with.) Leaves oblong or lanceolate, clasping.—Wet places.

2. **M. Jamesii,** Torr. Stem creeping at the base. *Corolla yellow*, small. Leaves roundish or kidney-shaped, nearly sessile. Calyx inflated in fruit.—In cool springs.

8. GRATI'OLA, L. HEDGE-HYSSOP.

1. **G. Virginia'na,** L. Stem 4-6 inches high, clammy with minute pubescence above. *Sterile filaments minute or none.* Leaves lanceolate. Peduncles slender.—Moist places.

2. **G. au'rea,** Muhl. *Nearly glabrous.* Sterile filaments slender, tipped with a little head. *Corolla golden yellow.*

9. ILYSAN'THES, Raf. FALSE PIMPERNEL.

I. **gratioloi'des,** Benth. Stem 4-8 inches high, much branched, diffusely spreading. Leaves ovate, rounded or oblong, the upper partly clasping.—Wet places.

10. GERAR'DIA, L. GERARDIA.

1. **G. purpu'rea,** L. (PURPLE GERARDIA.) Corolla rose-purple. Leaves linear, acute, rough-margined. *Flowers an inch long, on short peduncles.*—Low grounds.

2. **G. tenuifo'lia,** Vahl. (SLENDER G.) Corolla rose-purple. Leaves linear, acute. *Flowers about ½ an inch long, on long thread-like peduncles.*—Dry woods.

3. **G. fla'va,** L. (DOWNY G.) Corolla yellow, woolly inside. Stem 3-4 feet high, *finely pubescent.* Leaves oblong or lance-shaped, the upper entire, the lower usually more or less pinnatifid, downy-pubescent.—Woods.

4. **G. quercifo'lia,** Pursh. (SMOOTH G.) Corolla yellow, woolly inside. Stem 3-6 feet high, *smooth and glaucous.* Lower leaves twice-pinnatifid, the upper pinnatifid or entire, smooth.—Woods.

5. **G. pedicula'ria,** L. (CUT-LEAVED G.) Nearly smooth. Flowers nearly as in Nos. 2 and 4. Stem 2-3 feet high, very leafy, much branched. Leaves pinnatifid, *the lobes cut and toothed.*—Thickets.

11. CASTILLE'IA, Mutis. PAINTED-CUP.

C. coccin'ea, Spreng. (SCARLET PAINTED-CUP.) Calyx 2-cleft, yellowish. Stem pubescent or hairy, 1-2 feet high. The stem-leaves nearest the flowers 3-cleft, the lobes toothed, *bright scarlet.* (A yellow-bracted form occurs on the shore of Lake Huron.) —Sandy soil.

12. **EUPHRA'SIA,** Tourn.   Eyebright.

E. officina'lis, L., is rather common on the Lower St. Lawrence and the sea-coast. Lowest leaves crenate, those next the flowers *bristly-toothed.*

13. **RHINAN'THUS,** L.   Yellow-Rattle.

R. Crista-galli, L. (Common Yellow-Rattle.) Localities much the same as those of Euphrasia. Seeds broadly winged, rattling in the inflated calyx when ripe.

14. **PEDICULA'RIS,** Tourn.   Lousewort.

1. **P. Canadensis,** L. (Common Lousewort. Wood Betony.) Stems clustered, simple, hairy. Lowest leaves pinnately-parted. Flowers in a short spike.—Copses and banks.

2. **P. lanceola'ta,** Michx., has a nearly simple, smooth, upright stem, and oblong-lanceolate cut-toothed leaves. Calyx 2-lobed, leafy-crested. Pod ovate.—Grassy swamps.

15. **MELAMPY'RUM,** Tourn.   Cow-Wheat.

M. America'num, Michx. Leaves lanceolate, short-petioled; the lower ones entire.—Open woods.

Order LIX. **VERBENA'CEÆ.** (Vervain Family.)

Herbs (with us), with opposite leaves, didynamous stamens, and corolla either irregularly 5-lobed or 2-lipped. Ovary in Verbena 4-celled (when ripe splitting into 4 nutlets) and in Phryma 1-celled, *but in no case 4-lobed,* thus distinguishing the plants of this Order from those of the next.

Synopsis of the Genera.

1. **Verbe'na.** Flowers in spikes. Calyx tubular, 5-ribbed. Corolla tubular, salver-form, the border rather irregularly 5-cleft. Fruit splitting into 4 nutlets.
2. **Phry'ma.** Flowers in loose slender spikes, *reflexed in fruit.* Calyx cylindrical, 2-lipped, the upper lip of three slender teeth. Corolla 2-lipped. Ovary 1-celled and 1-seeded.

1. VERBE'NA, L. VERVAIN.

1. **V. hasta'ta,** L. (BLUE VERVAIN.) Stem 3-5 feet high. Leaves oblong-lanceolate, taper-pointed, serrate. Spikes of *purple* flowers dense, erect, corymbed or panicled.—Low meadows and fields.

2. **V. urticifo'lia,** L. (NETTLE-LEAVED V.) Stem tall. Leaves oblong-ovate, acute, coarsely serrate. Spikes of small *white* flowers very slender, loosely panicled.—Fields and roadsides.

3. **V. angustifo'lia,** Michx. Stem *low*. Leaves narrowly lanceolate, tapering at the base, sessile, roughish, slightly toothed. Flowers purple, in a crowded spike.—Dry soil.

2. PHRY'MA, L. LOPSEED.

P. Leptostach'ya, L. Corolla purplish or pale rose-coloured. Stem slender and branching, 1-2 feet high. Leaves ovate, coarsely toothed.—Woods and thickets.

ORDER LX. LABIA'TÆ. (MINT FAMILY.)

Herbs with square stems, opposite leaves (mostly aromatic), didynamous (or in one or two genera *diandrous*) stamens, a 2-lipped or irregularly 4- or 5-lobed corolla, and a *deeply 4-lobed ovary*, forming in fruit 4 nutlets or achenes. (See Part I., Section 65, for description of a typical plant.)

Synopsis of the Genera.

* *Stamens 4, curved upwards, parallel, exserted from a deep notch on the upper side of the 5-lobed corolla.*

1. **Teu'crium.** Calyx 5-toothed. The four upper lobes of the corolla nearly equal, *with a deep notch between the upper 2*; the lower lobe much larger. Flowers pale purple.

2. **Isan'thus.** Calyx bell-shaped, 5-cleft, almost equalling the small pale-blue corolla. Lobes of the corolla *almost equally spreading*. Stamens only slightly exserted.

* * *Stamens 4, the outer or lower pair longer, or only 2 with anthers, straight and not converging in pairs! Anthers 2-celled !*

+ Corolla almost equally 4-lobed, quite small.

3. **Men'tha.** Calyx equally 5-toothed. Upper lobe of the corolla rather the broadest, and sometimes notched. *Stamens 4*, of equal length, not convergent. Flowers either in terminal spikes or in head-like whorled clusters, often forming interrupted spikes. Corolla purplish or whitish.

LABIATÆ.   101

1. **Lycopus.** Calyx-teeth 4 or 5. *Stamens 2*, the upper pair, if any, without anthers. Flowers white, in dense axillary clusters.

+ + *Corolla evidently 2-lipped, but the lobes nearly equal in size; the tube not bearded inside. Stamens with anthers 2.*

5. **Hedeoma.** Calyx 2-lipped, bulging on the lower side at the base, hairy in the throat; 2 stamens with good anthers, *and 2 sterile filaments with false anthers.* Low odorous plants, with bluish flowers in loose axillary clusters.

+ + + *Corolla 2-lipped, the lower of the 5 lobes much larger than the other 4; the tube with a bearded ring inside. Stamens 2 (occasionally 4), much exserted.*

6. **Collinsonia.** Calyx ovate, enlarged and turned down in fruit, 2-lipped. Corolla elongated, the lower lip toothed or fringed. Strong-scented plants with *yellowish flowers on slender pedicels* in terminal panicled racemes.

+ + + + *Corolla evidently 2-lipped. Stamens with anthers 4.*

7. **Pycnanthemum.** Calyx short-tubular, 10–13-nerved, equally 5-toothed. The whitish or purplish flowers in *small dense* heads, forming terminal corymbs. Aromatic plants, with narrow rigid leaves crowded and and clustered in the axils.

8. **Satureia.** Calyx bell-shaped, not hairy in the throat, equally 5-toothed. Aromatic plants, with narrow leaves and purplish spiked flowers.

* * * *Stamens only 2, parallel; the anthers only 1-celled. Corolla 2-lipped.*

9. **Monarda.** Calyx tubular, nearly equally 5-toothed, hairy in the throat. Corolla elongated, strongly 2-lipped, the upper lip narrow. Stamens with long protruding filaments, each bearing a linear anther on its apex. Flowers large, in whorled heads surrounded by bracts.

* * * * *Stamens 4, the upper or inner pair longer! Anthers approximate in pairs. Corolla 2-lipped.*

10. **Nepeta.** Calyx obliquely 5-toothed. Anthers approaching each other in pairs under the inner lip of the corolla, the cells of each anther divergent. (See Figs. 59 and 60, Part I.)

11. **Lophanthus.** Calyx obliquely 5-toothed. Stamens exserted, the upper pair declined, the lower ascending, so that *the pairs cross.* Anther-cells *nearly parallel.* Tall herbs with small flowers in interrupted terminal spikes.

12. **Calamintha.** Calyx tubular, 2-*lipped*, often bulging below. Corolla 2-lipped, *the upper lip not arched,* the throat inflated. Flowers pale purple, in globular more or less dense clusters which are crowded with linear or awl-shaped hairy bracts.

* * * * *Stamens 4, the lower or outer pair longer! Anthers approximate in pairs. Corolla 2-lipped.*

13. **Physostegia.** Calyx not 2-lipped, 5-toothed or lobed, thin and membranaceous, inflated-bell-shaped in fruit. Anther-cells parallel. Flowers

large and showy, rose-colour variegated with purple, opposite, in terminal leafless spikes.

14. **Brunel'la.** *Calyx 2-lipped*, flat on the upper side, closed in fruit; *the upper lip 3-toothed, the lower 2-cleft*. Filaments 2-toothed at the apex, the lower tooth bearing the anther. Flowers violet, in a close terminal spike or head which is very leafy-bracted.

15. **Scutella'ria.** *Calyx 2-lipped*, short, closed in fruit, the lips rounded and entire, *the upper with a projection on the back*. Corolla blue or violet, the tube elongated and somewhat curved. Anthers of the lower stamens 1-celled, of the upper 2-celled. Flowers solitary in the axils of the upper leaves, or in axillary or terminal 1-sided racemes.

16. **Marru'bium.** *Calyx 10-toothed*, the teeth spiny and recurved after flowering. Stamens 4, *included in the corolla tube. Whitish woolly plants* with small white flowers in head-like whorls.

17. **Galeop'sis.** Calyx 5-toothed, the teeth spiny. The middle lobe of the lower lip of the corolla inversely heart-shaped, *the palate with 2 teeth at the sinuses*. Stamens 4, *the anthers opening cross-wise*. Flowers purplish, in axillary whorls.

18. **Stach'ys.** Calyx 5-toothed, beset with stiff hairs, the teeth spiny, diverging in fruit. Stamens 4, *the outer pair turned down after discharging their pollen*. Flowers purple, crowded in whorls, these at length forming an interrupted spike.

19. **Leonu'rus.** Calyx 5-toothed, the teeth spiny, and spreading when old. The middle lobe of the lower lip of the corolla inversely heart-shaped. Flowers pale purple, in close whorls in the axils of the *cut-lobed leaves*.

1. **TEU'CRIUM**, L. GERMANDER.

T. Canadense, L. (AMERICAN GERMANDER. WOOD SAGE.) Stem 1-3 feet high, downy. Leaves ovate-lanceolate, serrate, short-petioled, hoary beneath. Flowers in a long spike.—Low grounds.

2. **ISAN'THUS**, Michx. FALSE PENNYROYAL.

I. cæru'leus, Michx. A low, branching, clammy-pubescent annual. Leaves lance-oblong, 3-nerved, nearly entire. Peduncles axillary, 1-3-flowered.—Gravelly soil.

3. **MEN'THA**, L. MINT.

1. **M. vir'idis**, L. (SPEARMINT.) *Flowers in a narrow terminal spike*. Leaves ovate-lanceolate, wrinkled-veiny, unequally serrate, *sessile*.—Wet places.

2. **M. piperi'ta**, L. (PEPPERMINT.) *Flowers in loose interrupted spikes*. Leaves ovate or ovate-oblong, acute, *petioled*. Plant smooth.—Wet places.

LABIATÆ.

3. **M. Canadensis,** L. (WILD MINT.) *Flowers in axillary whorled clusters,* the uppermost axils without flowers. Stem more or less hairy, with ovate or lanceolate toothed leaves on short petioles. Var. **glabra'ta,** Benth., is smoothish, and has a rather pleasanter odour.—Shady wet places.

4. LYC'OPUS, L. WATER HOREHOUND.

1. **L. Virgin'icus,** L. (BUGLE-WEED.) *Calyx-teeth 4, bluntish.* Stems obtusely 4-angled, 6-18 inches high, producing slender runners from the base. Leaves ovate-lanceolate, toothed.—Moist places.

2. **L. Europæ'us,** L., var. **sinua'tus,** Gray. *Calyx-teeth 5, sharp-pointed.* Stem sharply 4-angled, 1-3 feet high. Leaves varying from cut-toothed to pinnatifid.—Wet places.

5. HEDEO'MA, Pers. MOCK PENNYROYAL.

1. **H. pulegioi'des,** Pers. (AMERICAN PENNYROYAL.) Stem 5-8 inches high, branching, hairy. Leaves oblong-ovate, *petioled,* obscurely serrate. Whorls few-flowered. Plant with a pungent aromatic odour.—Open woods and fields.

2. **H. his'pida,** Pursh., has the leaves *sessile, linear, and entire,* and the calyx *ciliate and hispid.*—Not common.

6. COLLINSO'NIA, L. HORSE-BALM.

C. Canaden'sis, L. (RICH-WEED. STONE-ROOT.) Stem smooth or nearly so, 1-3 feet high. Leaves serrate, pointed, petioled, 3-6 inches long.—Rich woods.

7. PYCNANTHEMUM, Michx. MOUNTAIN MINT. BASIL.

P. lanceola'tum, Pursh. Stem 2 feet high, smoothish or minutely pubescent. Leaves lanceolate or lance-linear. Heads downy. Calyx-teeth short. Lips of the corolla very short.—Dry soil.

8. SATURE'IA, L. SAVORY.

S. horten'sis, L. (SUMMER SAVORY.) Stem pubescent. Clusters few-flowered.—Escaped from gardens in a few localities.

9. MONAR'DA, L. HORSE-MINT.

1. **M. did'yma,** L. (OSWEGO TEA.) *Corolla bright red,* very showy. The large outer bracts tinged with red.—Along shaded streams.

2. **M. fistulo′sa,** L. (WILD BERGAMOT.) *Corolla purplish.*
The outer bracts somewhat purplish.—Dry and rocky banks and woods.

10. NEP′ETA, L. CAT-MINT.

1. **N. Cata′ria,** L. (CATNIP.) Flowers in cymose clusters. Stem erect, downy, branching. Leaves oblong, crenate, whitish beneath. Corolla dotted with purple.—Roadsides.

2. **N. Glecho′ma,** Benth. (GROUND IVY.) *Creeping and trailing.* Leaves *round-kidney-shaped,* crenate, green both sides. Corolla light blue.—Damp waste grounds.

11. LOPHAN′THUS, Benth. GIANT HYSSOP.

1. **L. nepetoi′des,** Benth. Smooth or nearly so, coarsely crenate-toothed. Calyx-teeth ovate, rather obtuse. Corolla greenish-yellow.—Borders of woods.

2. **L. scrophulariæfo′lius,** Benth., has lanceolate calyx-teeth and a purplish corolla.—Near Queenston Heights.

12. CALAMIN′THA, Mœnch. CALAMINTH.

1. **C. Clinopo′dium,** Benth. (BASIL.) Stem hairy, erect, 1-2 feet high. Flower-clusters *dense.* Leaves ovate, nearly entire, petioled.—Thickets and waste places.

2. **C. Nuttallii,** Benth. Smooth, 5-9 inches high. Leaves narrowly oblong. Clusters *few-flowered,* the flowers on slender naked pedicels. Bracts linear or oblong.—Wet limestone rocks, western and south-western Ontario.

13. PHYSOSTE′GIA, Benth. FALSE DRAGON-HEAD.

P. Virginia′na, Benth. Stem smooth, wand-like. Lower leaves oblong-ovate, upper lanceolate. Corolla an inch long, funnel-form, the throat inflated; upper lip slightly arching, the lower 3-parted, spreading, small.—Wet banks; not common.

14. BRUNEL′LA, Tourn. SELF-HEAL.

B. vulga′ris, L. (COMMON HEAL-ALL.) A low plant with oblong-ovate petioled leaves. Clusters 3-flowered, the whole forming a close terminal elongated head. Woods and fields everywhere.

BORRAGINACEÆ.

15. **SCUTELLA'RIA,** L. Skull-cap.

1. **S. galericula'ta,** L. Flowers blue, ¾ of an inch long, solitary in the axils of the upper leaves. Stem nearly smooth, 1-2 feet high.—Wet places.

2. **S. par'vula,** Michx. Flowers blue, ¼ of an inch long, solitary in the upper axils. Stem *minutely downy*, 3-6 inches high. Lowest leaves round-ovate, the upper narrower, all *entire.*—Dry banks.

3. **S. lateriflo'ra,** L. Flowers blue, ⅛ of an inch long, in 1-sided racemes. Stem upright, much branched, 1-2 feet high.— Wet places.

16. **MARRU'BIUM,** L. Horehound.

M. vulga're, L. Leaves round-ovate, crenate-toothed. Calyx with 5 long and 5 short teeth, recurved.—Escaped from gardens in some places.

17. **GALEOP'SIS,** L. Hemp-Nettle.

G. Tetra'hit, L. (Common Hemp-Nettle.) Stem bristly-hairy, swollen below the joints. Leaves ovate, coarsely serrate. Corolla often with a purple spot on the lower lip.—Waste places and fields.

18. **STACHYS,** L. Hedge-Nettle.

S. palustris, L., var. **as'pera,** Gray. Stem 2-3 feet high, 4-angled, the angles beset with stiff reflexed hairs or bristles.— Wet grounds.

19. **LEONU'RUS,** L. Motherwort.

L. Cardi'aca, L. (Common Motherwort.) Stem tall. Leaves long-petioled, the lower palmately lobed, the upper 3-cleft. Upper lip of the corolla bearded.—Near dwellings.

Order LXI. **BORRAGINA'CEÆ.** (Borage Family.)

Herbs, with a deeply 4-lobed ovary, forming 4 seed-like nutlets, as in the last Order, *but the corolla is regularly 5-lobed, with 5 stamens inserted upon its tube.*

Synopsis of the Genera.

* *Corolla without any scales in the throat.*

1. **E'chium.** Corolla with a funnel-form tube and a spreading border of 5 *somewhat unequal lobes. Stamens exserted, unequal.* Flowers bright blue with a purplish tinge, in racemed clusters. Plant bristly.

COMMON CANADIAN WILD PLANTS.

* * *Corolla with 5 scales completely closing the throat.*

2. **Sym'phytum.** Corolla tubular-funnel form with short spreading lobes *scales awl-shaped. Flowers yellowish-white*, in nodding raceme-like clusters, the latter often in pairs. *Nutlets smooth.* Coarse hairy herbs.

3. **Echinosper'mum.** *Nutlets prickly on the margin.* Corolla salver-shaped, the lobes rounded ; scales short and blunt. *Flowers blue*, small, in leafy-bracted racemes. Plant rough-hairy.

4. **Cynoglos'sum.** *Nutlets prickly all over.* Corolla funnel-form ; scales blunt. Flowers red-purple or pale blue, in racemes which are naked above, but usually leafy-bracted below. Strong-scented coarse herbs.

* * * *Corolla open, the scales or folds not sufficient to completely close the throat. Nutlets smooth.*

5. **Onosmo'dium.** Corolla tubular, *the 5 lobes acute and erect or converging.* Anthers mucronate ; filaments very short. Style thread-form, much exserted. Flowers greenish or yellowish-white. Rather tall stout plants, shaggy with spreading bristly hairs, or rough with short appressed bristles.

6. **Lithosper'mum.** Corolla funnel-form or salver-shaped, the 5 lobes of the spreading limb rounded. Anthers almost sessile. Root mostly red. Flowers small and almost white, or large and deep yellow, scattered or spiked and leafy-bracted.

7. **Myoso'tis.** Corolla salver-shaped, with a very short tube, *the lobes convolute in the bud;* scales or appendages of the throat blunt and arching. Flowers blue, in (so-called) racemes without bracts. Low plants, mostly soft-hairy.

1. **E'CHIUM**, Tourn. VIPER'S BUGLOSS.

E. vulga're, L. (BLUE-WEED.) Stem erect, 2 feet high. Leaves sessile, linear-lanceolate. Flowers showy, in lateral clusters, the whole forming a long narrow raceme.—Roadsides ; common in eastern Ontario, and rapidly spreading westward.

2. **SYM'PHYTUM**, Tourn. COMFREY.

S. officina'le, L. (COMMON COMFREY.) Stem winged above by the decurrent bases of the leaves, branched. Leaves ovate-lanceolate or lanceolate.—Moist soil ; escaped from gardens.

3. **ECHINOSPER'MUM**, Swartz. STICKSEED.

E. Lap'pula, L. Lehm. A very common roadside weed, 1-2 feet high, branching above. Leaves lanceolate, rough. Nutlets warty on the back, with a double row of prickles on the margin.

4. **CYNOGLOS'SUM**, Tourn. HOUND'S-TONGUE.

1. **C. officina'le**, L. (COMMON HOUND'S-TONGUE.) *Flowers red-purple.* Upper leaves lanceolate, sessile. Stem soft-pubescent. Nutlets rather flat.—A common weed in fields and along roadsides.

2. **C. Virgin'icum**, L. (WILD COMFREY.) *Flowers pale blue.* Stem roughish with spreading hairs. *Leaves few*, lanceolate-oblong, *clasping*. Racemes corymbed, raised on a long naked peduncle.—Rich woods.

3. **C. Moriso'ni**, DC. (*Echinospermum Virginicum*, Lehm., in Macoun's Catalogue.) (BEGGAR'S LICE.) *Flowers pale blue or white.* Stem hairy, *leafy*, with broadly spreading branches. Leaves taper-pointed and tapering at the base. Racemes panicled, forking, widely spreading. Pedicels of the flowers reflexed in fruit.—Open woods and thickets.

5. **ONOSMO'DIUM**, Michx. FALSE GROMWELL.

1. **O. Carolinia'num**, DC. Stem stout, 3-4 feet high. Leaves ovate-lanceolate, acute. *Lobes of the corolla ovate-triangular, very hairy outside.*—Banks of streams.

2. **O. Virginia'num**, DC. Stem slender, 1-2 feet high. Leaves narrowly oblong. *Lobes of the corolla lance-awl-shaped, sparingly bearded outside with long bristles.*—Banks and hillsides; not common.

6. **LITHOSPER'MUM**, Tourn. GROMWELL. PUCCOON.

* *Corolla almost white. Nutlets wrinkled, gray.*

1. **L. arven'se**, L. (CORN GROMWELL.) Stem 6-12 inches high, erect. Leaves lanceolate or linear.—Sandy banks.

** *Corolla deep yellow. Nutlets smooth and shining.*

2. **L. hirtum**, Lehm. (HAIRY PUCCOON.) Stem 1-2 feet high, *hispid*. Stem-leaves lanceolate or linear; those of the flowering branches ovate-oblong, ciliate. Flowers *peduncled. Corolla woolly at the base inside.*—Dry woods.

3. **L. canes'cens**, Lehm. (HOARY PUCCOON. ALKANET.) Stem 6-15 inches high, *soft-hairy. Corolla naked at the base inside.* Flowers *sessile.* Limb of the corolla smaller, and the calyx shorter, than in No. 2.—Open woods and plains.

*** *Corolla greenish-white or cream colour. Nutlets smooth and shining, mostly white.*

4. **L. officina'le**, L. (COMMON GROMWELL.) Much branched above. Leaves broadly lanceolate, acute. *Corolla exceeding the calyx.*—Roadsides and old fields.

5. **L. latifo'lium**, Michx. Loosely branched above. Leaves ovate and ovate-lanceolate, mostly taper-pointed. *Corolla shorter than the calyx.*—Borders of woods.

7. MYOSO'TIS, L. FORGET-ME-NOT.

1. **M. palustris**, Withering, var. **laxa**, Gray. (*Myosotis laxa*, Lehm., in Macoun's Catalogue.) (FORGET-ME-NOT.) Stem ascending from a creeping base, about a foot high, *smoothish*, loosely branched. *Calyx open in fruit.* Corolla pale blue, with a yellow eye. Pedicels spreading.—Wet places.

2. **M. arvensis**, Hoffm. Stem erect or ascending, *hirsute*. *Calyx closing in fruit.* Corolla blue, rarely white. Pedicels spreading in fruit and longer than the 5-cleft equal calyx. *Racemes naked at the base.*—Fields.

3. **M. verna**, Nutt., differs from the last in having a very small *white* corolla, pedicels *erect* in fruit, and the racemes leafy at the base. The calyx, also, is unequally 5-toothed and hispid. -Dry hills.

ORDER LXII. **HYDROPHYLLA'CEÆ.** (WATERLEAF F.)

Herbs, with alternate cut-toothed or lobed leaves, and regular pentamerous and pentandrous flowers very much like those of the last Order, *but having a 1-celled ovary with the seeds on the walls (parietal).* Style 2-cleft. Flowers mostly in 1-sided cymes which uncoil from the apex. The only common Genus is

HYDROPHYL'LUM, L. WATERLEAF.

1. **H. Virgin'icum**, L. Corolla bell-shaped, the 5 lobes convolute in the bud; the tube with 5 folds down the inside, one opposite each lobe. *Stamens and style exserted, the filaments bearded below.* Stem *smoothish*. Leaves *pinnately* cleft into 5 7 divisions, the latter ovate-lanceolate, pointed, cut-toothed. *Calyx-lobes very narrow, bristly-ciliate.* Flowers white or pale

blue. Peduncles *longer* than the petioles of the upper leaves. Rootstocks scaly-toothed.—Moist woods.

2. **H. Canadense**, L., differs from the last in having the leaves *palmately* 5-7-lobed, and *rounded*; the peduncles *shorter* than the petioles; and the calyx-lobes *nearly smooth*.–Rich woods.

3. **H. appendicula'tum**, Michx. Stem, pedicels, and calyx *hairy*. Stem-leaves palmately 5-lobed and rounded, the lowest leaves pinnately divided. *Calyx with a small reflexed appendage in each sinus*. Stamens sometimes not exserted.—Rich woods.

ORDER LXIII. **POLEMONIA'CEÆ.** (POLEMONIUM F.)

Herbs, with regular pentamerous and pentandrous flowers, *but a 3-celled ovary and 3-lobed style. Lobes of the corolla convolute in the bud.* Calyx persistent. Represented commonly with us by only one Genus,

PHLOX, L. PHLOX.

1. **P. divarica'ta**, L. Corolla salver-shaped, with a long tube. Stamens short, *unequally inserted*. Stem ascending from a prostrate base, somewhat clammy. Leaves *oblong-ovate*. Flowers lilac or bluish, in a spreading loosely-flowered cyme. *Lobes of the corolla mostly obcordate.*—Moist rocky woods.

2. **P. pilo'sa**, L. Leaves *lanceolate or linear*, tapering to a sharp point. Lobes of the pink-purple corolla obovate, *entire*.— Southwestern Ontario.

3. **P. subula'ta**, L., the Moss Pink of the gardens, has escaped from cultivation in some places. Stem creeping and tufted in broad mats. Flowers mostly rose-colour.—Dry grounds.

ORDER LXIV. **CONVOLVULA'CEÆ.** (CONVOLVULUS F.)

Chiefly twining or trailing herbs, with alternate leaves and regular flowers. Sepals 5, imbricated. Corolla 5-plaited or 5-lobed and convolute in the bud. Stamens 5. Ovary 2-celled.

Synopsis of the Genera.

1. **Calyste'gia.** *Calyx enclosed in 2 large leafy bracts.* Corolla funnel-form, the border obscurely lobed. Pod 4-seeded.
2. **Convol'vulus.** *Calyx without bracts.*

2. **Cus'cuta.** *Leafless parasitic slender twiners*, with yellowish or reddish stems, attaching themselves to the bark of other plants. Flowers small, mostly white, clustered. Corolla bell-shaped. *Stamens with a fringed appendage at their base.*

1. CALYSTE'GIA, R. Br. BRACTED BINDWEED.

1. **C. se'pium,** R. Br. (*Convolvulus sepium*, L., in Macoun's Catalogue.) (HEDGE BINDWEED.) *Stem mostly twining.* Leaves halberd-shaped. Peduncles 4-angled. Corolla commonly rose-coloured.—Moist banks.

2. **C. spithamæ'a,** Pursh. (*Convolvulus spithamæus*, L., in Macoun's Catalogue.) Stem low and simple, upright or ascending, *not twining*, 6-12 inches high. Leaves oblong, more or less heart-shaped at the base. Corolla white.—Dry soil.

2. CONVOL'VULUS, L. BINDWEED.

C. arvensis, L. (BINDWEED.) Stem twining or procumbent, and low. Leaves ovate-oblong, sagittate, the lobes acute. Corolla white, or tinged with red.

3. CUS'CUTA, Tourn. DODDER.

C. Grono'vii, Willd. Stems resembling coarse threads, spreading themselves over herbs and low bushes.—Low and moist shady places.

ORDER LXV. **SOLANA'CEÆ.** (NIGHTSHADE FAMILY.)

Rank-scented herbs (or one species shrubby), with colourless bitter juice, alternate leaves, and regular pentamerous and pentandrous flowers, *but a 2-celled ovary, with the placentæ in the axis.* Fruit a many-seeded berry or pod.

Synopsis of the Genera.

1. **Sola'num.** Corolla wheel-shaped, 5-lobed, the margins turned inward in the bud. *Anthers conniving around the style,* the cells opening by pores at the apex; filaments very short. The larger leaves often with an accompanying smaller one. Fruit a *berry.*
2. **Phys'alis.** Calyx 5-cleft, enlarging after flowering, *becoming at length much inflated, and enclosing the berry.* Corolla between wheel-shaped and funnel-form. *Anthers separate,* opening lengthwise. Plant clammy-pubescent.

SOLANACEÆ. 111

2. **Lycium.** Corolla funnel-form or tubular. Fruit a small *berry*, the calyx persistent but *not inflated*. A shrubby plant with long drooping branches and greenish-purple flowers on slender peduncles fascicled in the axils.
4. **Hyoscy'amus.** *Fruit a pod, the top coming off like a lid.* Calyx urn-shaped, 5-lobed, persistent. Corolla funnel-form, *oblique*, the limb 5-lobed, dull-coloured and veiny. Plant clammy-pubescent.
5. **Datu'ra.** Fruit a large prickly naked *pod.* Calyx long, *5-angled*, not persistent. Corolla very large, funnel-form, *strongly plaited* in the bud, with 5 pointed lobes. Stigma 2-lipped. Rank-scented weeds, with the showy flowers in the forks of the branching stems.
6. **Nicotia'na.** Fruit a *pod*, enclosed in the calyx. Calyx tubular-bell-shaped, 5-cleft. Corolla dull greenish-yellow, funnel-form, plaited in the bud. Leaves large. Flowers racemed or panicled.

1. SOLA'NUM, Tourn. NIGHTSHADE.

1. **S. Dulcama'ra, L.** (BITTERSWEET.) Stem somewhat shrubby and *climbing*. Leaves ovate and heart-shaped, *the upper halberd-shaped, or with 2 ear-like lobes at the base.* Flowers violet-purple, in small cymes. *Berries red.*—Near dwellings and in moist grounds.

2. **S. nigrum, L.** (COMMON NIGHTSHADE.) *Stem low and spreading*, branched. *Leaves ovate, wavy-toothed.* Flowers small, white, drooping in umbel-like lateral clusters. *Berries black.*—Fields and damp grounds.

3. **S. rostra'tum,** Dunal, is a *prickly* herb with large yellow flowers and *sharp* anthers.—Ottawa.

2. PHYS'ALIS, L. GROUND CHERRY.

1. **P. visco'sa, L.** (*P. Virginiana,* Mill, in Macoun's Catalogue.) Corolla *greenish-yellow,* brownish in the centre. Anthers yellow. Leaves ovate or heart-shaped, mostly toothed. Berry orange, sticky.—Sandy soil.

2. **P. grandiflo'ra,** Hook. Corolla *white,* large, with a woolly ring in the throat. Anthers tinged with blue or violet.

3. LYCIUM, L. MATRIMONY-VINE.

L. vulga're, Dunal. Common about dwellings. Berry oval, orange-red.

4. HYOSCY'AMUS, Tourn. HENBANE.

H. niger, L. (BLACK HENBANE.) Escaped from gardens in some localities. Corolla dull yellowish, netted with purple veins.

Leaves clasping, sinuate-toothed. A strong-scented and poisonous herb.

5. DATU'RA, L. STRAMONIUM. THORN-APPLE.

1. **D. Stramo'nium,** L. (COMMON THORN-APPLE.) Stem *green*. Corolla *white*, 3 inches long. Leaves ovate, sinuate-toothed.—Roadsides.

2. **D. Ta'tula,** L. (PURPLE T.) Stem *purple*. Corolla *pale violet-purple*.

6. NICOTIA'NA, L. TOBACCO.

N. rus'tica, L. (WILD TOBACCO.) Old fields and in gardens.

ORDER LXVI. GENTIANA'CEÆ. (GENTIAN FAMILY.)

Smooth herbs, *distinguished by having a 1-celled ovary with seeds on the walls, either in 2 lines or on the whole inner surface*. Leaves mostly opposite, simple, and sessile, but in one Genus alternate and compound. Stamens as many as the lobes of the regular corolla and alternate with them. Stigmas 2. Calyx persistent. Juice colourless and bitter.

Synopsis of the Genera.

1. **Fra'sera.** Corolla wheel-shaped, 4-parted ; a fringed glandular spot on each lobe. Flowers light greenish-yellow, with small purple-brown spots.
2. **Hale'nia.** Corolla 4-lobed, *the lobes all spurred at the base*. Flowers yellowish or purplish, somewhat cymose.
3. **Gentia'na.** Corolla not spurred, 4-5-lobed, mostly funnel-form or bell-shaped, generally with teeth or folds in the sinuses of the lobes. Stigmas 2, persistent. Pod oblong. Seeds innumerable. Flowers showy, in late summer and autumn.
4. **Menyan'thes.** A bog-plant. Corolla short funnel-form, 5-lobed, *densely white-bearded on the upper face. Leaves alternate, compound, of 3 oval leaflets. The flowers in a raceme at the summit of a naked scape*, white or tinged with pink.
5. **Limnan'themum.** An aquatic, with simple round-heart-shaped floating leaves on long petioles. Corolla white, wheel-shaped, 5-parted, bearded at the base only. Flowers in an umbel borne on the petiole.

1. FRA'SERA, Walt. AMERICAN COLUMBO.

F. Carolinien'sis, Walt. Tall and showy. Leaves whorled, mostly in fours. Root thick. Flowers numerous in a pyramidal panicle.—Dry soil.

2. **HALE′NIA**, Borkh. SPURRED GENTIAN.

II. **deflex′a**, Griseb. Stem erect, 9-18 inches high. Leaves 3-5-nerved, those at the base of the stem oblong-spathulate, petioled; the upper acute and sessile or nearly so. Spurs of the corolla curved.—Not common in Ontario; common on the Lower St. Lawrence.

3. **GENTIA′NA, L.** GENTIAN.

1. **G. crini′ta**, Fræl. (FRINGED GENTIAN.) Corolla funnel-form, *4-lobed, the lobes fringed on the margins;* no plaited folds in the sinuses. *Flowers sky-blue, solitary on long naked stalks* terminating the stem or simple branches. *Ovary lanceolate.* Leaves lance-shaped or ovate-lanceolate.—Low grounds.

2. **G. deton′sa**, Fries., (SMALLER FRINGED G.) is distinguished from No. 1 by the shorter or almost inconspicuous fringe of the corolla, the linear or lance-linear leaves, and the broader ovary. —Moist grounds, chiefly in the Niagara District.

3. **G. quinqueflo′ra**, Lam. (FIVE-FLOWERED G.) Corolla tubular-funnel-form, pale-blue, no folds in the sinuses. Calyx 5-cleft, the lobes awl-shaped. Lobes of the corolla triangular-ovate, bristle-pointed. Anthers separate. Stem slender and branching, a foot high, the branches racemed or panicled, about 5-flowered at the summit.—Dry hill-sides.

4. **G. alba**, Muhl. (WHITISH G.) Corolla inflated-club-shaped, *at length open,* 5-lobed, the lobes about twice as long as the *toothed appendages in the sinuses.* Flowers *greenish-white or yellowish,* sessile, crowded in a terminal cluster. Anthers usually cohering. Leaves lance-ovate, with a clasping heart-shaped base.—Low grounds.

5. **G. Andrews′ii**, Griseb. (CLOSED G.) Corolla inflated-club-shaped, *closed at the mouth,* the apparent lobes being really the large fringed-toothed appendages. *Flowers blue,* in a close sessile terminal cluster. Anthers cohering. Leaves ovate-lanceolate from a narrower base.—Low grounds; common northward, flowering later than No. 3.

4. **MENYAN′THES**, Tourn. BUCKBEAN.

M. trifolia′ta, L. A common plant in bogs and wet places northward. The bases of the long petioles sheathe the lower part of the scape, or thick rootstock, from which they spring. Plant about a foot high.

5. **LIMNAN'THEMUM**, Gmelin. FLOATING HEART.

L. lacuno'sum, Griseb. In shallow waters, northern Ontario.

ORDER LXVII. **APOCYNA'CEÆ.** (DOGBANE FAMILY.)

Herbs or slightly shrubby plants, with milky juice, opposite simple entire leaves, and regular pentamerous and pentandrous flowers with the lobes of the corolla convolute in the bud. *Distinguished by having 2 separate ovaries,* but the 2 stigmas united. Calyx free from the ovaries. Anthers converging round the stigmas. Seeds with a tuft of down on the apex. Represented with us only by the Genus

APO'CYNUM, Tourn. DOGBANE.

1. **A. androsæmifo'lium,** L. (SPREADING DOGBANE.) The corolla bell-shaped, 5-cleft, *pale rose-coloured, the lobes turned back. Branches of the stem widely forking.* Flowers in loose rather spreading cymes. Leaves ovate, petioled. Fruit 2 long and slender diverging pods.—Banks and thickets.

2. **A. cannab'inum,** L. (INDIAN HEMP.) Lobes of the *greenish-white* corolla not turned back. *Branches erect.* Cymes closer than in No. 1, and the flowers much smaller.—Along streams.

ORDER LXVIII. **ASCLEPIADA'CEÆ.** (MILKWEED F.)

Herbs with milky juice and opposite or whorled simple entire leaves. Pods, seeds, and anthers as in the last Order, *but the anthers are more closely connected with the stigma, the (reflexed) lobes of the corolla are valvate in the bud, the pollen is in waxy masses, and the (monadelphous) short filaments bear 5 curious hooded bodies behind the anthers.* Flowers in umbels. Commonly represented by only one Genus, which is typical of the whole Order.

ASCLE'PIAS, L. MILKWEED.

1. **A. Cornu'ti,** Decaisne. (COMMON MILKWEED.) Stem tall and stout. Leaves oval or oblong, short-petioled, pale green, 4–8

inches long. *Flowers dull greenish-purple. Pods ovate, soft spiny, woolly.*—Mostly in dry soil; very common.

2. **A. phytolaccoi'des,** Pursh. (POKE-MILKWEED.) Stem tall and smooth. Leaves broadly ovate, acute at both ends, short-petioled. *Pedicels loose and nodding, very long and slender.* Corolla greenish, *with the hooded appendages white.* Pods minutely downy, *but not warty.*—Moist thickets.

3. **A. incarna'ta,** L. (SWAMP M.) Stem tall, leafy, branching, and smooth. Leaves oblong-lanceolate, acute, obscurely heart-shaped at the base. *Flowers rose-purple. Pods very smooth and glabrous.*—Swamps and low grounds.

4. **A. tubero'sa,** L. (BUTTERFLY-WEED. PLEURISY-ROOT.) Stem very leafy, branching above, rough-hairy. Leaves linear or oblong-lanceolate, chiefly scattered. Corolla greenish-orange, *with the hoods bright orange-red.* Pods hoary. Dry hill-sides and fields; almost destitute of milky juice.

ORDER LXIX. **OLEA'CEÆ.** (OLIVE FAMILY.)

The only common representative Genus of this Order in Canada is Fraxinus (Ash). The species of this Genus are trees with pinnate leaves, and polygamous or diœcious flowers without petals and mostly also without a calyx; stamens only 2, with large oblong anthers. Fruit a 1-2-seeded samara. Flowers insignificant, from the axils of the previous year's leaves.

FRAX'INUS, Tourn. ASH.

1. **F. America'na,** L. (WHITE ASH.) *Fruit winged from the apex only, the base cylindrical. Branchlets and petioles smooth and glabrous.* Calyx very minute, persistent. Leaflets 7-9, *stalked.*—Rich woods.

2. **F. pubes'cens,** Lam., (RED ASH) has the *branchlets and petioles softly-pubescent,* and the fruit acute at the base, 2-edged, and gradually expanding into the long wing above.—Same localities as in No. 1.

3. **F. sambucifo'lia,** Lam. (BLACK or WATER ASH.) Branchlets and petioles smooth. Leaflets 7-9, *sessile,* serrate. *Fruit winged all round.* Calyx wanting, and the flowers consequently naked.—Swamps.

III. APET'ALOUS DIVISION.

Flowers destitute of corolla and sometimes also of calyx.

ORDER LXX. **ARISTOLOCHIA'CEÆ.** (BIRTHWORT F.)

Herbs with perfect flowers, *the tube of the 3-lobed calyx adherent to the 6-celled many-seeded ovary.* Leaves heart-shaped or kidney-shaped, on long petioles from a thick rootstock. Stamens 12 or 6. Flowers solitary. Calyx dull-coloured, the lobes valvate in the bud.

AS'ARUM, Tourn. WILD GINGER.

A. Canadense, L. Radiating stigmas 6. Leaves only a single pair, kidney-shaped, and rather velvety, the peduncle in the fork between the petioles, close to the ground. Rootstock aromatic. Calyx brown-purple inside, the spreading lobes pointed. —Rich woods.

ORDER LXXI. **PHYTOLACCA'CEÆ.** (POKEWEED F.)

Herbs with alternate leaves and perfect flowers, resembling in most respects the plants of the next Order, *but the ovary is composed of several carpels in a ring, forming a berry in fruit.* Only one Genus and one Species.

PHYTOLAC'CA, Tourn. POKEWEED.

P. decan'dra, L. (COMMON POKE.) Calyx of 5 rounded white sepals. Ovary green, of 10 1-seeded carpels united in a ring. Styles 10, short and separate. Stamens 10. Fruit a crimson or purple 10-seeded berry. Stem very tall and stout, smooth. Flowers in long racemes opposite the leaves.—Sandy soil.

ORDER LXXII. **CHENOPODIA'CEÆ.** (GOOSEFOOT F.)

Homely herbs, with more or less succulent leaves (chiefly alternate), and small greenish flowers mostly in interrupted spikes. Stamens usually as many as the lobes of the calyx and opposite them. Ovary 1-celled and 1-ovuled, forming an achene or utricle in fruit. Stigmas mostly 2.

Synopsis of the Genera.

1. **Chenopo'dium.** Weeds with (usually) mealy leaves, and very small perfect greenish sessile flowers in small panicled spiked clusters. Calyx 5-cleft, more or less enveloping the fruit. Stamens mostly 5; filaments slender.
2. **Bli'tum.** Flowers perfect, in heads which form interrupted spikes. Calyx becoming fleshy and bright red in fruit so that the clusters look something like strawberries. Leaves triangular and somewhat halberd-shaped, sinuate-toothed.
3. **At'riplex** Flowers *monœcious or diœcious*, the staminate with a regular calyx in spiked clusters; the pistillate without a calyx, but with a pair of appressed bracts.
4. **Corisper'mum.** Flowers all *perfect, single*, and sessile in the axils of the upper leaves, usually forming a spike. *Calyx of a single delicate sepal.* Low herbs, with linear 1-nerved leaves.
5. **Salso'la**, with fleshy awl-shaped sharp-pointed leaves, is not uncommon on the Lower St. Lawrence and the sea-coast.

1. CHENOPO'DIUM, L. GOOSEFOOT. PIGWEED.

1. **C. album, L.** (LAMB'S-QUARTERS.) Stem upright, 1–3 feet high. Leaves varying from rhombic-ovate to lanceolate, more or less toothed, *mealy, as are also the dense flower-clusters.*—Extremely common in cultivated soil.

2. **C. ur'bicum, L.** Rather pale and only slightly mealy, 1–3 feet high, branches *erect*. Leaves triangular, acute, *coarsely and sharply many-toothed.* Spikes erect, crowded in a long and narrow racemose panicle.—Waste places in towns.

3. **C. hy'bridum, L.** (MAPLE-LEAVED GOOSEFOOT.) *Bright green.* Stem widely branching, 2–4 feet high. Leaves thin, large, triangular, heart-shaped, sinuate-angled, the angles extended into pointed teeth. Panicles loose, leafless. Plant with a rank unpleasant odour.—Waste places.

4. **C. Bo'trys, L.** (JERUSALEM OAK.) Not mealy, but *sticky;* low, spreading, sweet-scented. Leaves deeply sinuate, *slender-petioled.* Racemes in *loose* divergent corymbs. — Roadsides; escaped from gardens.

5. **C. ambrosioi'des, L.** (MEXICAN TEA.) Not mealy, but sticky. Leaves *slightly* petioled, wavy-toothed or nearly entire. Spikes *densely* flowered.—Streets of towns.

2. BLI'TUM, Tourn. BLITE.

B. capita'tum, L. (STRAWBERRY BLITE.) Stem ascending, branching. Leaves smooth. The axillary head-like clusters very conspicuous in fruit.—Dry soil, margins of woods, &c.

3. AT'RIPLEX, Tourn. ORACHE.

A. pat'ula, L. Erect or diffuse, scurfy, green or rather hoary. Leaves varying from triangular or halberd-shaped to lance-linear, petioled.—Streets of towns.

4. CORISPER'MUM, Ant. Juss. BUG-SEED.

C. hyssopifo'lium, L. Somewhat hairy when young, pale. Stamens 1 or 2. Styles 2. Fruit oval, flat.—Sandy beaches, western and south-western Ontario.

ORDER LXXIII. **AMARANTA'CEÆ.** (AMARANTH F.)

Homely weeds, a good deal like the plants of the last Order, but the flower-clusters are interspersed with dry and chaff-like (sometimes coloured) persistent bracts, usually 3 to each flower.

Synopsis of the Genera.

1. **Amaran'tus.** Flowers *monœcious or polygamous*, all with a calyx of 3 or 5 distinct erect sepals.
2. **Monte'lia.** Flowers *diœcious;* calyx none in the pistillate flowers.

1. AMARAN'TUS, Tourn. AMARANTH.

1. A. panicula'tus, L. *Reddish flowers* in terminal and axillary *slender* spikes, the bracts awn-pointed.—In the neighbourhood of gardens.

2. A. retroflex'us, L. (PIGWEED.) *Flowers greenish,* in spikes forming a stiff panicle. Leaves a dull green, long-petioled, ovate, wavy-margined. *Stem erect.*—Common in cultivated soil.

3. A. albus, L. Flowers greenish, in small close axillary clusters. *Stem low and spreading.*—Roadsides.

2. MONTE'LIA, Moquin.

M. tamaris'cina, Gray. A tall smooth herb, with lanceolate or oblong-ovate alternate leaves on long petioles, and small clusters of greenish flowers in interrupted spikes.—Wet places.

ORDER LXXIV. **POLYGONA'CEÆ.** (BUCKWHEAT F.)

Herbs, *well marked by the stipules of the alternate leaves being in the form of membranous sheaths above the usually swollen joints of the stem.* Flowers usually perfect. Calyx 4-6-parted. Stamens 4-9, inserted on the base of the calyx. Stigmas 2 or 3. Ovary 1-celled, with a single ovule rising from the base, forming a little nutlet.

Synopsis of the Genera.

1. **Polyg'onum.** Sepals 5 (occasionally 4), often coloured and petal-like, *persistent*, embracing the 3-angled (or sometimes flattish) nutlet or achene. Flowers in racemes or spikes, or sometimes in the axils.
2. **Ru'mex.** Sepals 6, *the 3 outer ones herbaceous and spreading in fruit*, the 3 inner (called *valves*) somewhat petal-like and, after flowering, convergent over the 3-angled achene, *often with a grain-like projection on the back*. Stamens 6. Styles 3. Flowers usually in crowded whorls, the latter in panicled racemes.
3. **Fagopy'rum.** Calyx 5-parted, petal-like. Stamens 8, *with 8 yellow glands between them.* Styles 3. Achene 3-angled. Flowers white, in panicles. Leaves triangular heart-shaped or halberd-shaped.

1. POLYG'ONUM, L. KNOTWEED.

* *Flowers along the stem, inconspicuous, greenish-white, nearly sessile in the axils of the small leaves. Sheaths cut-fringed or torn.*

1. **P. avicula're,** L. (KNOTGRASS. GOOSEGRASS.) A weed everywhere in yards and waste places. *Stem prostrate and spreading.* Stamens chiefly 5. Achene 3-sided, *dull*. Stigmas 3. Leaves sessile, lanceolate or oblong. Var. **erectum** is upright and larger, with broader leaves.

2. **P. ten'ue,** Michx. Stem *slender, upright,* sparingly branched. Leaves sessile, *narrowly linear,* very acute. Achene *smooth and shining.*—Dry soil and rocky places.

* * *Flowers in terminal spikes or racemes, mostly rose-coloured or pinkish, occasionally greenish.*

← *Leaves not heart-shaped or arrow-shaped.*

3. **P. incarna'tum,** Ell. *Sheaths not fringed. Stem nearly smooth,* 3-6 feet high. Leaves long, tapering from near the base to a narrow point, rough on the midrib and margins. *Spikes linear and nodding.* Stamens 6. Styles 2. *Achene flat or hollow-sided.*—In muddy places along streams and ponds.

4. **P Pennsylva'nicum,** L. Sheaths not fringed. Stem 1-3 feet high, *the upper branches and the peduncles bristly with stalked glands.* Spikes thick, erect. Stamens 8. Achene flat.—Low open grounds.

5. **P. Persica'ria,** L. (LADY'S THUMB.) Sheaths with a somewhat ciliate border. *Stem nearly smooth,* a foot or more in height. *Leaves with a dark blotch on the middle of the upper surface.* Spikes dense, erect, on naked peduncles. Stamens 6. Achene flat or 3-angled, according as the stigmas are 2 or 3.—Very common near dwellings in moist ground.

6. **P. amphib'ium,** L. (WATER PERSICARIA.) Spike of flowers dense, oblong, *showy, rose-red. Stem floating in shallow water* or rooting in soft mud, *nearly glabrous.* Leaves long-petioled, *often floating.* Sheaths not bristly-fringed. *Stamens 5.* Stigmas 2.—In shallow water, mostly northward.

7. **P. Muhlenberg'ii,** Watson, differs from the last in being *rough with appressed hairs* all over.—Ditches.

8. **P. Hartwright'ii,** Gray, is distinguished from P. amphibium by its *foliaceous and ciliate sheaths.*—Muddy margins of ponds and lakes.

9. **P hydropiperoi'des,** Michx. (MILD WATER-PEPPER.) Stem slender, 1-3 feet high, *in shallow water.* Leaves narrow, roughish. *Sheaths hairy and fringed with long bristles.* Spikes slender, erect, pale rose-coloured or whitish. Stamens 8. Stigmas 3. Achene 3-angled.—In shallow water.

10. **P. acre,** H.B.K. (WATER SMARTWEED.) Sheaths fringed with bristles. *Leaves transparent-dotted.* Stem rooting at the decumbent base, 2-4 feet high, *in shallow water or muddy soil.* Leaves narrow, taper-pointed. Spikes slender, *erect,* pale rose-coloured. *Sepals glandular-dotted.* Stamens 8. Achene 3-angled, *shining.*—Muddy soil or shallow water.

11. **P. Hydrop'iper,** L. (COMMON SMARTWEED or WATER-PEPPER.) Sheaths and leaves as in the last, the leaves, however, larger. *Spikes slender, nodding,* greenish. *Sepals glandular-dotted.* Stamens 6. Achene dull.—Wet places.

12. **P. Virginia'num,** L. Calyx greenish, *unequally 4-parted.* Stamens 5. Styles 2, persistent on the flat achene. Flowers in

long and slender naked spikes. Stem upright, nearly smooth. Leaves ovate or ovate-lanceolate, taper-pointed, rough ciliate. Sheaths hairy and fringed.—Thickets, in rich soil.

+ + *Leaves heart-shaped or sagittate. Sheaths much longer on one side than on the other.*

13. **P. arifo'lium, L.**, (HALBERD-LEAVED TEAR-THUMB) with grooved stem, halberd-shaped long-petioled leaves, flowers in short loose racemes, 6 stamens, and a flattish achene, is not uncommon on the Lower St. Lawrence; rare in Ontario.

14. **P. sagitta'tum, L.** (ARROW-LEAVED TEAR-THUMB.) *Stem 4-angled, the angles beset with reflexed minute prickles,* by which the plant is enabled to climb. Leaves arrow-shaped. Stamens 8. Achene 3-angled.—Common in low grounds, especially beaver-meadows.

15. **P. Convol'vulus, L.** (BLACK BINDWEED.) *Stem twining,* not prickly but *roughish; the joints naked.* Flowers in loose panicled racemes, 3 of the calyx-lobes ridged in fruit. Leaves heart-shaped and partly halberd-shaped Not climbing so high as the next.—Cultivated grounds and waste places.

16. **P. dumeto'rum, L.**, var. **scandens,** Gray. (CLIMBING FALSE BUCKWHEAT.) Stem twining high, *smooth; sheaths naked,* 3 of the calyx-lobes *winged in fruit.*—Moist thickets.

17. **P. cilino'de,** Michx. Stem twining, *minutely downy. Sheaths fringed at the base with reflexed bristles.*—Sandy pine woods and rocky hills.

2. **RUMEX,** L. Dock. SORREL.

* *Herbage not sour, nor the leaves halberd-shaped.*

1. **R. orbicula'tus,** Gray. (GREAT WATER DOCK.) Growing in marshes. Stem erect, stout, 5-6 feet high. Leaves lanceolate, *not wavy-margined or heart-shaped,* often over a foot long. Flowers nodding *on thread-like pedicels.* Valves nearly orbicular, finely net-veined, each with a grain on the back.—Wet places.

2. **R. salicifo'lius,** Weinmann, (WHITE DOCK) may be looked for in marshes on the sea-coast and far northward. The whorls of flowers are dense and form a very conspicuous spike, owing to the great size of the grains on the back of the valves.

3. **R. verticilla'tus,** L. (SWAMP DOCK.) Leaves lanceolate or oblong-lanceolate, *not wavy*, the *lowest* often heart-shaped. Stem tall. Fruit-bearing pedicels slender, club-shaped, abruptly reflexed, several times longer than the fruiting calyx. Valves dilated-rhomboid, *strongly wrinkled*, each bearing a very large grain.—Swamps, common.

4. **R. crispus,** L. (CURLED DOCK.) *Leaves with strongly wavy or curly margins*, lanceolate. Whorls of flowers in long wand-like racemes. Valves grain-bearing.—Cultivated soil and waste places.

5. **R. obtusifo'lius,** L. (BITTER DOCK.) Lowest leaves oblong heart-shaped, obtuse, only slightly wavy-margined; the upper oblong-lanceolate, acute. Whorls loose, *distant*. Valves somewhat halberd-shaped, *sharply toothed at the base*, usually *one only* grain-bearing.—Waste grounds.

** *Herbage sour ; leaves halberd-shaped.*

6. **R. Acetosel'la,** L (FIELD or SHEEP SORREL.) Stem 6-12 inches high. *Flowers diœcious*, in a terminal naked panicle.—A very common weed in poor soil.

3. FAGOPY'RUM, Tourn. BUCKWHEAT.

F. esculen'tum, Mœnch. (BUCKWHEAT.) Old fields and copses, remaining after cultivation.

ORDER LXXV. LAURA'CEÆ. (LAUREL FAMILY.)

Trees or shrubs with spicy-aromatic bark and leaves, the latter simple (often lobed), alternate, and marked with small transparent dots (visible under a lens). Sepals 6, petal-like. Flowers diœcious or polygamo-diœcious. Stamens in sterile flowers 9, inserted at the base of the calyx. Anthers opening by uplifting valves. Ovary in fertile flowers free from the calyx, 1-celled, with a single ovule hanging from the top of the cell. Style and stigma 1. Fruit a 1-seeded drupe.

1. SAS'SAFRAS, Nees. SASSAFRAS.

S. officina'le, Nees. A small or moderate-sized tree with yellowish or greenish-yellow twigs and ovate or 3-lobed entire leaves. Flowers greenish-yellow, in naked corymbs, appearing with the leaves in the axils of the latter. Drupe *blue*, on a reddish pedicel. The 9 stamens in 3 rows, the 3 inner each with a pair of yellow glands at the base of the filament. *Anthers 4-celled, 4-valved.*—Rich woods, in southern and western Ontario.

2. **LIN'DERA**, Thunberg. WILD ALLSPICE. FEVER-BUSH.

L. Benzo'in, Meisner. (SPICE-BUSH.) A nearly smooth shrub with oblong-obovate leaves, pale beneath. Flowers honey-yellow, in lateral umbel-like clusters, before the leaves. Stamens very much as in Sassafras, but the anthers are *2-celled and 2-valved.* Pistillate flowers with 15–18 rudiments of stamens. *Drupe red.* —Damp woods, in early spring.

ORDER LXXVI. **THYMELEA'CEÆ.** (MEZEREUM F.)

Shrubs with tough leather-like bark and entire leaves. Flowers perfect. Calyx tubular, resembling a corolla, pale yellow. Stamens twice as many as the lobes of the calyx (in our species 8). Style thread-like. Stigma capitate. Ovary 1-celled, 1-ovuled, free from the calyx. Fruit a berry-like drupe. Only one Species in Canada.

DIRCA, L. LEATHERWOOD. MOOSE-WOOD.

D. palustris, L. A branching shrub, 2–5 feet high, with curious *jointed branchlets* and nearly oval leaves on short petioles. Flowers in clusters of 3 or 4, preceding the leaves. Filaments exserted, half of them longer than the others.—Damp woods.

ORDER LXXVII. **ELÆAGNA'CEÆ.** (OLEASTER F.)

Shrubs with diœcious flowers, *and leaves which are scurfy on the under surface.* The calyx-tube in the fertile flowers *becomes fleshy and encloses the ovary, forming a berry-like fruit.* Otherwise the plants of this Order are not greatly different from those of the last.

SHEPHERD'IA, Nutt. SHEPHERDIA.

S. Canadensis, Nutt. Calyx in sterile flowers 4-parted. Stamens 8. Calyx in fertile flowers urn-shaped, 4-parted. Berries yellow. Branchlets brown-scurfy. Leaves opposite, entire, ovate, green above, silvery-scurfy beneath, the small flowers in their axils.—Gravelly banks of streams and lakes.

ORDER LXXVIII. **SANTALA'CEÆ.** (SANDALWOOD F.)

Low herbaceous or partly woody plants (with us) with perfect flowers, these *greenish-white*, in terminal or axillary corymbose clusters. Calyx bell-shaped or urn-shaped, 4–5-cleft, adherent to the 1-celled ovary, lined with a 5-lobed disk, the stamens on the edge of the latter between its lobes and *opposite the lobes of the calyx, to which the anthers are attached by a tuft of fine hairs.* Fruit nut-like, crowned with the persistent calyx-lobes.

COMAN'DRA, Nutt. BASTARD TOAD-FLAX.

C. umbella'ta, Nutt. Stem 8–10 inches high, leafy. Leaves oblong, pale green, an inch long. Flower-clusters at the summit of the stem. Calyx-tube prolonged and forming a neck to the fruit. Style slender.—Dry soil.

ORDER LXXIX. **SAURURA'CEÆ.** (LIZARD'S-TAIL F.)

A small family having, with us, but a single representative:—

SAURU'RUS, L. LIZARD'S-TAIL.

S. cer'nuus, L. A swamp herb, with jointed branching stem, 2 feet high. Leaves petioled, heart-shaped, with converging ribs. Flowers white, in a dense terminal spike, nodding at the end, each flower with a lanceolate bract. Flowers perfect, but entirely destitute of calyx and corolla. Stamens usually 6 or 7, with long slender white filaments. Carpels 3 or 4, slightly united at the base.

ORDER LXXX. **CERATOPHYLLA'CEÆ.** (HORNWORT F.)

Represented, with us, by a single species.

CERATOPHYL'LUM, L. HORNWORT.

C. demer'sum, L. An aquatic herb, with whorled finely dissected leaves, and minute axillary sessile monœcious flowers, without calyx or corolla, but with an 8–12-cleft involucre. Staminate flowers of 12–24 stamens with large sessile anthers. Pistillate flowers of a single 1-celled ovary, forming an achene, beaked with the slender style. *Embryo with 4 cotyledons.*—Under water in ponds and slow streams.

EUPHORBIACEÆ. 125

ORDER LXXXI. **EUPHORBIA'CEÆ.** (SPURGE F.)

Plants with milky juice and monœcious flowers, represented in Canada chiefly by the two following genera:

1. EUPHOR'BIA, L. SPURGE.

Flowers monœcious, the sterile and fertile ones both destitute of calyx and corolla, *but both contained in the same 4-5-lobed cup-shaped involucre which resembles a calyx,* and therefore the whole will probably at first sight be taken for a single flower. Sterile flowers numerous, *each of a single naked stamen from the axil of a minute bract.* Fertile flower only 1 in each involucre; ovary 3-lobed, *soon protruded on a long pedicel;* styles 3, each 2-cleft. Peduncles terminal, often umbellate.

* *Leaves all similar and opposite, short-petioled, green or blotched with brown above, furnished with scale-like or fringed stipules. Stems spreading or prostrate, much forked. Involucres in terminal or lateral clusters, or one involucre in each fork, the involucre invariably with 4 (mostly petal-like) glands in the sinuses.*

1. **E. polygonifo'lia,** L. Leaves *entire,* oblong-linear, mucronate, *very smooth.* Stipules bristly-fringed. Peduncles in the forks. *Glands of the involucre very small, not petal-like.* Pods obtusely angled.—Shores of the Great Lakes, in sandy or gravelly places.

2. **E. serpens,** H. B. K. Leaves *entire,* round-ovate, very small, smooth. Stipules membranaceous, triangular. Peduncles longer than the petioles, in loose clusters. Glands of the small involucre with minute crenulate appendages. Stems thread-like, prostrate. Pods acutely angled. *Seeds smooth.*—London and westward, not common.

3. **E. glyptosper'ma,** Engel. Leaves *serrulate* towards the apex, linear-oblong, *very unequal at the base.* Stipules lanceolate, cut into bristles. Peduncles as long as the petioles, in dense lateral clusters. Glands of the small involucre with crenulate appendages. Stems erect-spreading. Pods sharply angled. *Seeds sharply 4-angled, with 5 or 6 sharp transverse wrinkles.*—Gravelly soil.

4. **E. macula'ta, L.** Leaves *serrulate*, oblong-linear, *somewhat pubescent, with a brownish blotch in the centre*, very oblique at the base. *Peduncles in dense lateral clusters.* Glands of the involucre with reddish petal-like attachments. *Pods sharply angled.*—Roadsides.

5. **E. hypericifo'lia, L.** Stem ascending. Leaves *serrate*, often with a red spot or with red margins, oblique at the base, ovate-oblong or oblong-linear. *Peduncles in cymes at the ends of the branches.* Glands of the involucre with *white* or occasionally reddish petal-like attachments. Pod smooth, obtusely angled.—Cultivated soil.

Only the uppermost or floral leaves whorled or opposite. Stems erect. Stipules none. Involucres 5-lobed; inflorescence umbelliform, in the forks of the branches, and terminal.

6. **E. corolla'ta, L.** Conspicuous for the *5 bright-white false lobes of the involucre, resembling petals*; the true lobes very small.—Gravelly or sandy soil.

* * * Involucres chiefly in terminal umbels, and their glands always without petal-like attachments. Leaves without stipules or blotches, those of the stem alternate or scattered, the floral ones usually of a different shape, and whorled or opposite.

7. **E. platyphyl'la, L.** Umbel 5-rayed. Stem erect, 8-18 inches high. Upper stem-leaves lance-oblong, acute, serrulate, the uppermost heart-shaped, the floral ones triangular-ovate and cordate. *Pod warty*—Shores of the Great Lakes.

8. **E. Heliosco'pia, L.** Umbel first 5-rayed, then with 3, and finally merely forked. Stem ascending, 6-12 inches high. Leaves all obovate, rounded or notched at the apex, serrate. *Pods smooth.*—Along the Great Lakes.

9. **E. Cyparis'sias, L.**, with densely clustered stems, and crowded linear stem-leaves (the floral ones round heart-shaped), and a many-rayed umbel, has escaped from gardens in some localities.

2. **ACALY'PHA, L.** THREE-SEEDED MERCURY.

A. Virgin'ica, L. Flowers monœcious, both kinds having a calyx, the staminate 4-parted, the pistillate 3-5-parted; no involucre. Staminate flowers very small, in spikes, with 1-3 pistillate flowers at the base, in the axil of a large leaf-like 5-9 lobed bract. Stamens 8–16, monadelphous at the base, the anther-cells hanging from the apex of the filament. Styles 3, the stigmas cut-fringed, usually red. Pod separating into 3 globular carpels. A nettle-like weed, with ovate, sparsely serrate, alternate, long-petioled leaves.—Fields and open places.

Order LXXXII. URTICA'CEÆ. (Nettle F.)

Herbs, shrubs, or trees, with monœcious or diœcious (or, in the Elms, sometimes perfect) flowers, with a regular calyx free from the 1–2-celled ovary which becomes a 1-seeded fruit. Stamens opposite the lobes of the calyx. This Order is divided into four well-marked Suborders.

Suborder I. ULMA'CEÆ. (Elm Family.)

Trees, with alternate simple leaves, and deciduous small stipules. Flowers often perfect. Styles 2. *Fruit a samara winged all round, or a drupe.*

* *Fruit a samara; anthers extrorse.*

1. **Ulmus.** Flowers in lateral clusters, earlier than the leaves, purplish or greenish-yellow. Calyx bell-shaped, 4-9-cleft. Stamens 4-9; the filaments long and slender. Ovary 2-celled, but the samara only 1-seeded. Stigmas 2.

** *Fruit a drupe; anthers introrse.*

2. **Celtis.** Flowers greenish, polygamous, the pistillate solitary or in pairs, appearing with the leaves. Calyx 5–6-parted, persistent. Stamens 5–6. Stigmas 2, long and pointed and recurved. Ovary 1-ovuled.

Suborder II. ARTOCAR'PEÆ. (Bread-fruit & Fig F.)

Flowers monœcious or diœcious, crowded in catkin-like spikes or heads, the whole pistillate catkin becoming an aggregate fruit from the enlargement of the calyx in the several flowers. Calyx 4-parted. Stamens 4. Ovary 2-celled, 1 cell eventually disappearing. Styles 2.

3. **Morus.** Pistillate and staminate flowers in separate catkins. Trees with milky juice and rounded leaves. Staminate spikes slender.

Suborder III. URTI'CEÆ. (Nettle Family.)

Herbs with watery juice and opposite or alternate leaves, often beset with stinging hairs. Flowers monœcious or diœcious, in spikes or racemes. Stamens as many as the sepals. Style only 1. Ovary 1-celled. Fruit an achene.

4. **Urti'ca.** *Leaves opposite.* Plant beset with stinging hairs. Sepals 4 in both sterile and fertile flowers. Stamens 4. Stigma a small sessile tuft. Achene flat, enclosed between the 2 larger sepals. Flowers greenish.

5. **Laport'ea.** *Leaves alternate.* Plant beset with stinging hairs. Sepals 5 in the sterile flowers, 4 in the fertile, 2 of them much smaller than the other 2. Stigma awl-shaped. Achene flat, *very oblique, reflexed on its winged pedicel.*

6. **Pil'ea.** Leaves opposite. *Whole plant very smooth and semi-transparent.* Sepals and stamens 3-4. Stigma a sessile tuft.

7. **Bœhme'ria.** Leaves mostly opposite. No stinging hairs. Sepals and stamens 4 in the sterile flowers. Calyx tubular or urn-shaped in the fertile ones, and enclosing the achene. Stigma long and thread-like.

8. **Parieta'ria.** Leaves alternate, entire, 3-ribbed. No stinging hairs. Flowers polygamous, in involucrate-bracted cymose axillary clusters. Calyx of the pistillate flowers tubular or bell-shaped, 4-lobed. Stigma tufted. Staminate flowers nearly as in the last.

SUBORDER IV. **CANNABIN'EÆ.** (HEMP FAMILY.)

Rough herbs with watery juice and tough bark. Leaves opposite and *palmately compound.* Flowers diœcious. Sterile ones in compound racemes; stamens 5; sepals 5. Fertile ones in crowded clusters; sepal only 1, embracing the achene. Stigmas 2.

9. **Can'nabis.** A rather tall rough plant with palmately compound leaves of 5-7 linear-lanceolate serrate leaflets.

1. **ULMUS,** L. ELM.

1. **U. fulva,** Michx. (RED or SLIPPERY ELM.) Flowers nearly sessile. *Leaves very rough above, taper-pointed. Buds downy with rusty hairs.* A medium-sized tree, with mucilaginous inner bark.

2. **U. America'na,** L. (AMERICAN or WHITE ELM.) *Leaves not rough above, abruptly pointed.* Flowers on drooping pedicels. *Buds glabrous.* A large ornamental tree, with drooping branchlets.—Moist woods.

3. **U. racemo'sa,** Thomas. (CORKY WHITE ELM.) Resembling the last, but the *bud-scales are downy-ciliate, the branches corky, and the flowers racemed.*—Chiefly along roadsides and borders of fields.

URTICACEÆ.

2. CEL'TIS, L.   Nettle-tree.   Hackberry.

C. occidenta'lis, L. (Sugarberry.) A small tree of Elm-like aspect. *Leaves reticulated*, ovate, taper-pointed, serrate, more or less oblique at the base. Fruit as large as a pea, dark-purple when ripe, the flesh thin.—Low grounds; a few trees here and there through Ontario.

3. MORUS, Tourn.   Mulberry.

M. rubra, L. (Red Mulberry.) Leaves heart-ovate, rough above, downy beneath, pointed. Fruit red, turning dark-purple, long.—Niagara district, and south-westward.

4. URTI'CA, Tourn.   Nettle.

U. gra'cilis, Ait. Stem slender, 2-6 feet high. Leaves ovate-lanceolate, pointed, serrate, 3-5-nerved from the base, nearly smooth, the long petioles with a few bristles. Flower-clusters in slender spikes.—Moist ground and along fences.

5. LAPORT'EA, Gaudichaud.   Wood-Nettle.

L. Canadensis, Gaudichaud. Stem 2-3 feet high. Leaves large, ovate, long-petioled, *a single 2-cleft stipule in the axil.*—Moist woods.

6. PIL'EA, Lindl.   Richweed.   Clearweed.

P. pu'mila, Gray. Stem 3-18 inches high. Leaves ovate, coarsely toothed, 3-ribbed.—Cool moist places.

7. BŒHME'RIA, Jacq.   False Nettle.

B. cylin'drica, Willd. Stem 1-3 feet high, smoothish. Leaves ovate-oblong or ovate-lanceolate, serrate, 3-nerved, long-petioled. Stipules separate.—Moist shady places.

8. PARIETA'RIA, Tourn.   Pellitory.

P. Pennsylvan'ica, Muhl. A low annual, simple or sparingly branched, minutely downy. Leaves oblong-lanceolate, thin, veiny, roughish with opaque dots.—Usually in crevices of limestone rocks; not very common.

9. CAN'NABIS, Tourn.   Hemp.

C. sati'va, L. (Hemp.) Common everywhere along roadsides and in waste places.

ORDER LXXXIII. **PLATANA'CEÆ.** (PLANE-TREE F.)

Represented only by the Genus

PLAT'ANUS, L. PLANE-TREE. BUTTONWOOD.

P. occidenta'lis, L. (AMERICAN PLANE-TREE or SYCAMORE.) A fine large tree found in south-western Ontario. Leaves alternate, rather scurfy when young, palmately lobed or angled, the lobes sharp-pointed; *stipules sheathing*. Flowers monœcious, both sterile and fertile ones in catkin-like heads, without calyx or corolla, but with small scales intermixed. Ovaries in the fertile flowers club-shaped, tipped with the thread-like simple style, and downy at the base. Fertile heads solitary, on slender peduncles. The white bark separates into thin plates.

ORDER LXXXIV. **JUGLANDA'CEÆ.** (WALNUT F.)

Trees with alternate pinnate leaves and no stipules. Flowers monœcious. Sterile flowers in catkins. Fertile flowers solitary or in small clusters, with a regular 3-4-lobed calyx adherent to the ovary. Fruit a sort of drupe, the fleshy outer layer at length becoming dry and forming a husk, the inner layer hard and bony and forming the nut-shell. Seed solitary in the fruit, very large and 4-lobed. This Order comprises the Walnuts, Butternuts, and Hickories.

Synopsis of the Genera.

1. **Jug'lans.** Sterile flowers in *solitary* catkins from the previous year's wood. Filaments of the numerous stamens very short. Fertile flowers on peduncles at the ends of the branches. Calyx 4-toothed, *with small petals at the sinuses*. Styles and stigmas 2, the latter fringed. *Exocarp or husk drying without splitting. Shell of the nut very rough and irregularly furrowed.*

2. **Car'ya.** Sterile flowers in slender *clustered* catkins. Stamens 3-10, with very short filaments. Fertile flowers in small clusters at the ends of the branches. Calyx 4-toothed; *no petals*. Stigmas 2 or 4, large. *Exocarp 4-valved, drying and splitting away from the very smooth and bony nut-shell.*

1. JUG'LANS, L. WALNUT.

1. J. ciner'ea, L. (BUTTERNUT.) Leaflets oblong-lanceolate, pointed, serrate. *Petioles and branchlets clammy. Fruit oblong, clammy.*—Rich woods.

2. **J. nigra,** L. (BLACK WALNUT.) Leaflets ovate-lanceolate, taper-pointed, serrate. Petioles downy *but not clammy*. Fruit spherical. Wood a darker brown than in the Butternut.—Rich woods; rare northward.

2. CAR'YA, Nutt. HICKORY.

1. **C. alba,** Nutt. (SHELL-BARK HICKORY.) *Leaflets 5*, the lower pair much smaller than the others. Husk of the fruit splitting *completely* into 4 valves. Nut flattish-globular, mucronate. Bark of the trunk rough, scaling off in rough strips.—Rich woods.

2. **C. tomento'sa,** Nutt. (WHITE-HEART HICKORY.) Sparingly found in the Niagara district and south-westward. Leaflets 7-9. Bark close but not shaggy, and not scaling off on the old trunks. Husk as in the last. Catkins, s..oots, and lower surface of the leaves *tomentose* when young. Nut *globular*.

3. **C. ama'ra,** Nutt. (SWAMP HICKORY or BITTERNUT.) *Leaflets 7-11*. Husk of the fruit splitting *half way* down. Nut *spherical*, short-pointed. Bark smooth, not scaling off.—Moist ground.

4. **C. porci'na,** Nutt. (PIG-NUT. BROOM-HICKORY.) Leaflets 5-7. Shoots, etc., *glabrous*. Husk as in the last. Nut *oblong or oval*.—Niagara district, and south-westward.

ORDER LXXXV. CUPULIF'ERÆ. (OAK FAMILY.)

Shrubs or trees, with alternate simple leaves, deciduous stipules, and monœcious flowers. Sterile flowers in catkins (but in Beech in small heads); the fertile ones solitary or clustered, and furnished with an involucre which forms a scaly cup or a bur surrounding the nut.

Synopsis of the Genera.

1. **Quercus.** Sterile flowers *with a calyx* including few or several stamens with slender filaments. Fertile flowers scattered or somewhat clustered, each in a scaly involucre or cupule. Nut (acorn) rounded, the base enclosed by the cupule. (Part I, sec. 71.)

2. **Casta'nea.** Sterile flowers in long slender catkins. *Calyx 6-parted.* Fertile flowers usually 3 in each involucre, the latter prickly, forming a bur. Calyx 6-lobed. Stigmas bristle-shaped. Nuts enclosed (mostly 2 or 3

together) in the prickly 4-valved involucre, flattened when there are more than one.

3. **Fagus.** Sterile flowers in a small head on drooping peduncles. *Calyx bell-shaped.* Fertile flowers in pairs in the involucre, which consists of awl-shaped bractlets grown together at the bases. Calyx-lobes awl-shaped. *Nuts 3-angled, generally in pairs in the bur-like 4-valved cupule.* Bark close, smooth, and light gray.

4. **Cor'ylus.** Sterile flowers in drooping catkins. *No calyx.* Stamens 8 (with 1-celled anthers), *and 2 small bractlets* under each bract. Fertile flowers in a small scaly head; one ovary, surmounted by 2 long red stigmas, under each scale, and accompanied by a pair of bractlets which, in fruit, enlarge and form a *leaf-like or tubular fringed or toothed involucre closely enveloping each nut.* Sterile catkins from the axils of the previous year. Fertile flowers terminating the new shoots.

5. **Os'trya.** Sterile flowers in drooping catkins. *Calyx wanting.* Stamens several under each bract, *but not accompanied by bractlets.* Fertile flowers in short catkins, 2 under each bract, each ovary tipped with 2 long stigmas, and *surrounded by a tubular bractlet which, in fruit, becomes a greenish-white inflated bag, having the small nut in the bottom.*

6. **Carpi'nus.** Sterile flowers in drooping catkins. *Calyx wanting.* Stamens several under each bract; *no bractlets.* Fertile flowers much as in Ostrya, *but the bractlets surrounding the ovaries are not tubular but open, and in fruit become leaf-like, one on each side of the small nut.*

1. QUERCUS, L. OAK.

* *Acorns ripening the first year, and therefore borne on the new shoots. Lobes or teeth of the leaves not bristle-pointed.*

1. **Q. alba,** L. (WHITE OAK.) A large tree. Leaves (when mature) smooth, bright green above, whitish beneath, obliquely cut into few or several oblong entire *lobes.* The oblong nut much larger than the saucer-shaped rough cupule.—Rich woods.

2. **Q. macrocar'pa,** Michx. (BUR OAK. MOSSY-CUP WHITE OAK.) A medium-sized tree. Leaves *deeply lobed,* smooth above, pale or downy beneath. Acorn broadly ovoid, *half or altogether covered by the deep cup, the upper scales of which taper into bristly points making a fringed border.* Cup varying greatly in size, often very large.—Rich soil.

3. **Q. bi'color,** Willd. (SWAMP WHITE OAK.) A tall tree. Leaves *sinuate-toothed,* but hardly lobed, *wedge-shaped at the base,* downy or hoary beneath, the main veins 6-8 pairs. Cup nearly hemispherical, about half as long as the oblong-ovoid acorn, some-

times with a fringed border. *Peduncle in fruit longer than the petiole.*—Low grounds.

4. **Q. Pri'nus,** L. (CHESTNUT OAK.) A small tree. Leaves minutely downy beneath, *the main veins 10-16 pairs,* sinuate-toothed, acute or obtuse at the base. *Peduncle shorter than the petiole.* Cup hemispherical; acorn as in the last.—Lake Erie coast.

Var. **hu'milis**, Marsh, (*Q. prinoides*, Willd., in Macoun's Catalogue) is much more abundant with us than the species itself. It has the characters of the species, but is a shrub, 2-4 feet high. Fruit sessile or nearly so.

** *Acorns ripening the second year, and therefore borne on the previous year's wood, below the leaves of the season. Lobes or teeth of the leaves bristle-pointed.*

5. **Q. coccin'ea,** Wang. (SCARLET OAK.) A large tree. Leaves bright green, shining above, turning red in autumn, rounded at the base, deeply pinnatifid, the lobes divergent and sparingly cut-toothed. Bark gray outside, *reddish* inside. *Cup top-shaped or hemispherical with a more or less conical base,* covering half or more of the rather small acorn.

Var. **tincto'ria,** Gray. (*Q. tinctoria,* Bartram, in Macoun's Catalogue.) (QUERCITRON, YELLOW-BARKED, or BLACK OAK.) Leaves usually less deeply pinnatifid, slender-petioled, rather rounded at the base, rusty-downy when young, smooth and shining above when mature, often slightly pubescent beneath, turning brownish, orange, or dull red in autumn. Cup as in the species, but the bark darker and rougher and *yellow or orange* inside.—Western Ontario; mostly in dry soil, but occasionally in moist places.

6. **Q. rubra,** L. (RED OAK.) A large tree. Leaves moderately pinnatifid, turning dark-red in the autumn. *Cup saucer-shaped,* sessile or nearly so, very much shorter than the oblong-ovoid acorn.—Rich and poor soil.

7. **Q. palustris,** Du Roi. (PIN OAK.) A medium-sized tree. Cup flat-saucer-shaped, very much shorter than the *ovoid-globose* acorn, which is about half an inch long. Leaves deeply pinnatifid, with divergent lobes and rounded sinuses.—Niagara district and south-westward.

2. **CASTA'NEA**, Tourn. CHESTNUT.

C. vesca, L., var. **America'na,** Michx. (*C. vulgaris,* var. *Americana,* A. DC., in Macoun's Catalogue.) (CHESTNUT.) A large tree. Leaves oblong-lanceolate, pointed coarsely and sharply serrate, acute at the base. Nuts 2 or 3 in each bur. —South-western Ontario.

3. **FAGUS**, Tourn. BEECH.

F. ferrugin'ea, Ait. (AMERICAN BEECH.) A very common tree in rich woods, the branches horizontal. Leaves oblong-ovate, taper-pointed, toothed, the very straight veins terminating in the teeth.

4. **COR'YLUS**, Tourn. HAZEL-NUT. FILBERT.

1. **C. America'na,** Walt. (WILD HAZEL-NUT.) Leaves roundish heart-shaped. *Involucre spreading out above, leaf-like and cut-toothed.*—Chiefly in south-western Ontario; in thickets.

2. **C. rostra'ta,** Ait. (BEAKED HAZEL-NUT.) A rather common shrub, easily distinguished from No. 1 by the involucre, *which is prolonged into a narrow tube much beyond the nut, and is densely bristly-hairy.*

5. **OS'TRYA**, Micheli. HOP-HORNBEAM. IRON-WOOD.

O. Virgin'ica, Willd. (IRON-WOOD.) A slender tree with brownish furrowed bark. Leaves oblong-ovate, taper-pointed, sharply doubly serrate. Fertile catkin like a hop in appearance. Wood very hard and close.—Rich woods.

6. **CARPI'NUS**, L. HORNBEAM.

C. America'na, Michx. (BLUE or WATER BEECH.) Small trees with *furrowed trunks* and close smooth gray bark. Leaves ovate-oblong, pointed, doubly serrate.—Along streams. Resembling a Beech in general aspect, but with inflorescence like that of Iron-wood.

ORDER LXXXVI. **MYRICA'CEÆ.** (SWEET-GALE F.)

Shrubs with monœcious or diœcious flowers, both sterile and fertile ones collected in short catkins or heads. *Leaves with resinous dots, usually fragrant.* Fruit a 1-seeded dry drupe or little nut, usually coated with waxy grains.

Synopsis of the Genera.

1. **Myri'ca.** Flowers chiefly diœcious, catkins lateral, each bract with a pair of bractlets underneath. Stamens in the sterile flowers 2-8. Ovary solitary in the fertile flowers, 1-celled, tipped with 2 thread-like stigmas, and surrounded by 2-4 small scales at the base. In our species the 2 scales form wings at the base of the nut.—A shrub, 3-5 feet high.

2. **Compto'nia.** A low shrub, a foot or more in height, with fern-like very sweet-scented leaves. Flowers monœcious. Sterile catkins cylindrical. Fertile ones spherical, the ovary surrounded by 8 awl-shaped persistent scales, so that the catkin resembles a bur.

1 MYRI'CA, L. BAYBERRY. WAX-MYRTLE.

M. Gale, L. (SWEET GALE.) Leaves wedge-lanceolate, serrate towards the apex, pale. The small nuts in crowded heads, and winged by the 2 scales.—Bogs.

2. COMPTO'NIA, Solander. SWEET-FERN.

C. asplenifo'lia, Ait. (*Myrica asplenifolia,* Endl., in Macoun's Catalogue.) Leaves linear-lanceolate in outline, deeply pinnatifid, the lobes numerous and rounded.—Dry soil; especially in Pine barrens.

ORDER LXXXVII. BETULA'CEÆ. (BIRCH FAMILY.)

Trees or shrubs with monœcious flowers, both sorts in catkins, 2 or 3 flowers under each scale or bract of the catkin. Ovary 2-celled and 2-ovuled, but in fruit only 1-celled and 1-seeded. Fruit a small nut. Stigmas 2, long and slender. Twigs and leaves often aromatic.

Synopsis of the Genera.

1. **Bet'ula.** Sterile catkins long and pendulous, formed during summer and expanding the following spring; each flower consisting of one small scale to which is attached 4 short filaments; 3 flowers under each scale of the catkin. Fertile catkins stout, oblong, the scales or bracts 3-lobed and with 2 or 3 flowers under each; each flower a naked ovary, becoming a winged nutlet in fruit. Bark easily coming off in sheets.

2. **Alnus.** Catkins much as in Betula, but each fertile and sterile flower has a distinct 3-5-parted calyx. Catkins solitary or clustered at the ends of leafless branchlets or peduncles. Nutlets wingless or nearly so.

(These two genera are included in Cupuliferæ in Macoun's Catalogue.)

1. BET'ULA, Tourn. BIRCH.

1. **B. lenta,** L. (CHERRY-BIRCH. SWEET or BLACK BIRCH.) Bark of the trunk dark brown, close, aromatic; that of the twigs bronze-coloured. Wood rose-coloured. Leaves ovate, with somewhat heart-shaped base, doubly serrate, pointed, short-petioled. Fruiting catkins sessile, thick, oblong-cylindrical.—Moist woods.

2. **B. lu'tea,** Michx. (YELLOW or GRAY BIRCH.) Bark of the trunk yellowish-gray, somewhat silvery, scaling off in thin layers. Leaves hardly at all heart-shaped. Fruiting catkins thicker and shorter than in No. 1.—Moist woods.

3. **B. papyra'cea,** Ait. (*B. papyrifera*, Michx., in Macoun's Catalogue.) (PAPER or CANOE BIRCH.) Bark of the trunk white, easily separating in sheets. Leaves ovate, taper-pointed, heart-shaped, long-petioled. Fruiting catkins cylindrical, usually hanging on slender peduncles.—Woods.

4. **B. pu'mila,** L. (LOW BIRCH.) A *shrub* with brownish bark, *not glandular*. Leaves obovate or roundish, *pale beneath;* veinlets on both surfaces finely reticulated. Catkins mostly erect, on short peduncles.—Bogs and low grounds, northward.

2. ALNUS, Tourn. ALDER.

1. **A. inca'na,** Willd. (SPECKLED or HOARY ALDER.) A shrub or small tree, growing in thickets in low grounds along streams. Leaves oval or ovate, rounded at the base, serrate, whitish beneath. Flowers preceding the leaves in early spring, from clustered catkins formed the previous summer and remaining naked over winter. *Fruit wingless.*

2. **A. vir'idis,** DC. (GREEN or MOUNTAIN ALDER.) A shrub 3-8 feet high, along mountain streams. Flowers appearing with the leaves, the *staminate* catkins having remained naked during the winter, *the pistillate enclosed in a scaly bud. Fruit with a thin wing.*—Northward.

ORDER LXXXVIII. SALICA'CEÆ. (WILLOW FAMILY.)

Trees or shrubs with diœcious flowers, both sorts in catkins, one under each scale of the catkin. No calyx. Fruit 1-celled, *many-seeded*, the seeds furnished with tufts of down. (See Part

I., section 74, for description of typical flowers.) This Order comprises the Willows and Poplars.

Synopsis of the Genera.

1. **Salix.** Trees with mostly *long and pointed leaves* and slender branches. Bracts or scales of the catkins *not toothed*. Stamens mostly 2 under each bract, but in one or two species as many as 5 or 6. Stigmas short. Catkins appearing before or with the leaves.

2. **Pop'ulus.** Trees with *broad* and more or less heart-shaped leaves. Bracts of the catkins *toothed or cut at the apex*. Stamens 8-30, or even more, under each scale. Stigmas long. Catkins long and drooping, preceding the leaves.

1. SALIX, Tourn. WILLOW.

* *Catkins lateral and sessile, appearing before the leaves. Scales dark red or brown, persistent. Usually no leaf-like bracts at the base of the catkins. Stamens 2.*

+ *Leaves veiny, hairy or woolly, and with somewhat revolute margins.*

1. **S. can'dida**, Willd. (HOARY WILLOW.) A shrub not more than 3 or 4 feet high, growing in bogs and wet places; the twigs and leaves clothed with a web-like wool, giving the whole plant a whitish aspect. Leaves lanceolate, narrow. Stipules small, lanceolate, toothed. Catkins cylindrical.

2. **S. hu'milis**, Marshall. (PRAIRIE WILLOW.) A shrub 3-8 feet high, growing usually in dry or barren places. Leaves lanceolate, not so taper-pointed as in No. 1, slightly downy above, thickly so beneath. Stipules semi-ovate or moon-shaped, with a few teeth, shorter than the petioles. Catkins ovoid.

+ + *Leaves smooth and shining above, not woolly beneath. Catkins large, clothed with long glossy hairs.*

3. **S. dis'color**, Muhl. (GLAUCOUS WILLOW.) A shrub or small tree, 8-15 feet high, growing in low grounds and along streams. Leaves lanceolate or ovate-lanceolate, irregularly toothed *in the middle of the margin*, entire at each end, white-glaucous beneath. Stipules moon-shaped, toothed.

The 3 species just described frequently have compact heads of leaves, resembling cones, at the ends of the branches. This is probably a diseased condition due to puncturing by insects.

4. **S. petiola'ris**, Smith. (PETIOLED WILLOW.) A shrub on sandy river-banks. Leaves lanceolate, finely and evenly serrate,

silky-gray or glaucous beneath, smooth above. Catkins with a few small leaf-like bracts at the base. Scales of the fertile catkins acute, very hairy. Ovary tapering, silky, stalked. Sandy river-banks.

* * *Catkins lateral, with 4 or 5 leafy bracts at the base, preceding (or sometimes accompanying) the leaves. Scales dark red or brown, persistent. Stamens 2.*

5. **S. corda'ta**, Muhl. (HEART-LEAVED WILLOW.) A shrub or small tree, growing in wet grounds. Leaves lanceolate, not always heart-shaped, sharply serrate, smooth. Catkins cylindrical, *leafy-bracted at the base.* Var. **angusta'ta** has long narrow leaves.

* * * *Catkins lateral, appearing along with the leaves, leafy-bracted at the base. Stamens 2. Scales persistent.*

6. **S. liv'ida**, Wahl. Var. **occidenta'lis**, Gray. (*S. rostra'ta,* Rich., in Macoun's Catalogue.) (LIVID WILLOW.) A good-sized shrub, chiefly in moist situations. *Leaves oblong* or obovate-lanceolate, barely toothed, downy above, very veiny, hairy and glaucous beneath. Stipules semi-lunar, toothed. Ovary at length raised on a very slender stalk.

* * * * *Catkins long and loose, peduncled, not lateral, but borne on the ends of the new shoots. Scales greenish-yellow, deciduous. Filaments hairy below.* ← *Stamens 3-6 or more.*

7. **S. lu'cida**, Muhl. (SHINING WILLOW.) A shrub or small bushy tree, growing along streams. *Leaves ovate-oblong or narrower,* with a long tapering point, *shining on both sides,* serrate. Stamens mostly 5.

8. **S. nigra**, Marshal. (BLACK WILLOW.) A larger tree than No. 6, with a roughish black bark, growing along streams. *Leaves narrowly lanceolate,* tapering at each end, serrate, smooth, green on both sides. Stamens 3-6.

← ← *Stamens 2.*

9. **S. longifo'lia**, Muhl. (LONG-LEAVED WILLOW.) A shrub or small tree, varying greatly in size, growing along streams in sandy or gravelly places. Leaves linear-lanceolate, very long, tapering towards both ends, nearly sessile, serrate with a few spreading teeth, grayish-hairy when young.

2. POP'ULUS, Tourn. POPLAR.

1. **P. tremuloi'des,** Michx. (AMERICAN ASPEN.) A tree with greenish-white bark, and *roundish heart-shaped leaves*, continually in a state of agitation, due to the lateral compression of the petiole, and the consequent susceptibility of the leaf to the least motion of the air. *Teeth of the leaves small.*

2. **P. grandidenta'ta,** Michx., (LARGE-TOOTHED ASPEN) has *roundish ovate leaves with large irregular sinuate teeth.*

3. **P. balsamif'era,** L. (BALSAM POPLAR.) A tall tree, growing in swamps and along streams; *the large buds varnished with resinous matter.* Leaves *ovate*, tapering, finely serrate, whitish beneath. Stamens very numerous.

4. **P. monilif'era,** Ait. (COTTONWOOD.) A tree with broad deltoid leaves, slightly heart-shaped, serrate with incurved teeth. Young branches *slightly angled*, at length round. Fertile catkins very long, the scales *cut-fringed, not hairy*. Along the main line of the Grand Trunk Railway.

Var. **can'dicans,** Gray, (BALM OF GILEAD) has broader and *more or less heart-shaped* leaves.

SUBCLASS II. GYM'NOSPERMS.

Ovules and seeds naked (not enclosed in a pericarp), and fertilized by the direct application of the pollen. Represented in Canada by a single Order.

ORDER LXXXIX. CONIF'ERÆ. (PINE FAMILY.)

Trees or shrubs with resinous juice and mostly monœcious flowers, these in catkins except in the last genus (Taxus), in which the fertile flower is solitary and the fruit berry-like. Leaves awl-shaped or needle-shaped. (See Part I., Cap. xvi., for descriptions of typical plants.)—The Order comprises three well-marked Suborders.

SUBORDER I. **ABIETIN'EÆ.** (PINE FAMILY PROPER.)

Fruit a true *cone*, the imbricated scales *in the axils of bracts*. Ovules 2 on the inside of each scale at the base, in fruit coming off with a wing attached to each. (Part I., Figs. 197, 198.)

* *Cones not ripening till the second year.*

1. **Pinus.** Leaves needle-shaped, *2-5 in a cluster, evergreen*, in the axil of a thin scale. Sterile catkins in spikes at the bases of the new shoots, consisting of many almost sessile anthers spirally inserted on the axis. Cones more or less woody, the scales widely spreading when ripe. Cotyledons of the embryo several.

** *Cones ripening the first year.*

2. **A'bies.** Leaves linear or needle-shaped, *scattered uniformly along the new shoots, evergreen.* Sterile catkins in the axils of last year's leaves. Cones with thin scales.

3. **Larix.** Leaves needle-shaped, *clustered or fascicled on lateral spurs of last year's wood, many in each bundle, falling off in the autumn;* those on the new shoots scattered, but *deciduous like the rest.*

SUBORDER II. **CUPRESSIN'EÆ.** (CYPRESS F.)

Fertile flowers of only a few scales, these *not in the axils of bracts*, forming in fruit either a very small loose and dry cone, or a sort of false berry owing to the thickening of the scales.

* *Flowers monœcious. Fruit a small loose cone.*

4. **Thuja.** Leaves some *awl-shaped*, others *scale-like*, closely imbricated on the *flat branches.* Catkins ovoid, terminal.

** *Flowers mostly diœcious. Fruit berry-like, black with a bloom.*

5. **Juniperus.** Leaves awl-shaped or scale-like, sometimes of both shapes, evergreen, prickly-pointed, glaucous-white on the upper surface, and in whorls of 3, or opposite.

SUBORDER III. **TAXIN'EÆ.** (YEW FAMILY.)

Fertile flower solitary, consisting of a naked ovule surrounded by a disk which becomes pulpy and berry-like in fruit, enclosing the nut-like seed. *Berry red.*

6. **Taxus.** Flowers chiefly diœcious. Leaves evergreen, *mucronate*, rigid, scattered.—A low straggling bush, usually in the shade o other evergreens.

1. **PINUS,** Tourn. PINE.

1. **P. resino'sa,** Ait. (RED PINE.) *Leaves in twos*, slender Bark rather smooth, *reddish.*—Common northward.

2. **P. stro'bus**, L. (WHITE PINE.) *Leaves in fives*, slender. Bark smooth except on old trees, not reddish.—Common.

2. A'BIES, Tourn. SPRUCE. FIR.

1. **A. nigra**, Poir. (*Picea nigra*, Link, in Macoun's Catalogue.) (BLACK SPRUCE.) *Leaves needle-shaped and 4-sided, pointing in all directions.* Cones hanging, *persistent*, scales with thin edges.—Swamps and cold woods.

2. **A. alba**, Michx. (*Picea alba*, Link, in Macoun's Catalogue.) (WHITE SPRUCE.) Leaves as in No. 1. Cones hanging, *deciduous*, the scales with thickish edges.—Swamps and cold woods.

3. **A. Canadensis**, Michx. (*Tsuga Canadensis*, Carr, in Macoun's Catalogue.) (HEMLOCK SPRUCE.) *Leaves flat*, lighter beneath, *pointing only in two directions, i.e.*, right and left on each side of the branch, obtuse. Cones hanging, persistent.—Hilly or rocky woods.

4. **A. balsa'mea**, Marshall. (BALSAM FIR.) *Leaves flat*, the lower surface whitish and the midrib prominent, crowded, pointing mostly right and left on the branches. *Cones erect* on the upper sides of the branches, violet-coloured, *the scales slender-pointed.*—Damp woods and swamps.

3. LARIX, Tourn. LARCH.

L. America'na, Michx. (AMERICAN LARCH. TAMARAC.) A slender and very graceful tree with soft leaves in fascicles, falling off in autumn.—Swamps.

4. THUJA, Tourn. ARBOR VITÆ.

T. occidenta'lis, L. (AMERICAN ARBOR VITÆ.) The well-known cedar of cedar-swamps.—Common.

5. JUNIP'ERUS, L. JUNIPER.

1. **J. commu'nis**, L. (COMMON JUNIPER.) A spreading shrub with ascending stems, growing on dry hill-sides. *Leaves in whorls of 3*, whitish above, prickly-pointed.

2. **J. Virgini'ana**, L. (RED CEDAR.) A shrub or small tree with *mostly opposite leaves* of two forms, viz.: awl-shaped and loose, and scale-shaped and appressed. Fruit small, erect. Wood reddish and odorous.—Dry sterile soil.

6. **TAXUS,** Tourn. YEW.

T. bacca'ta, L., var. **Canadensis,** Gray. (AMERICAN YEW. GROUND HEMLOCK.) A low straggling shrub. Leaves green on both sides. *Berry globular, red.*

Class II. MONOCOTYLE'DONS.

For characters of the Class see Part I., chap. xv.

I. SPADIC'EOUS DIVISION.

Flowers aggregated on a *spadix* (Part I., sec. 94), with or without a *spathe* or sheathing bract.

Order XC. **ARA'CEÆ.** (Arum Family.)

Herbs with pungent juice and simple or compound leaves, *these sometimes net-veined* and hence suggesting that the plants may be Dicotyledons. Spadix usually accompanied by a spathe. Flowers either without a perianth of any kind, or with 4–6 sepals. Fruit usually a berry.

Synopsis of the Genera.

* *Leaves not linear. Flowers without perianth of any sort. Spadix accompanied by a spathe.*

1. **Arisæ'ma.** Flowers mostly diœcious, collected on the lower part of the spadix only. Spathe (in our common species) arched over the spadix. Scape from a solid bulb. Leaves compound, net-veined, sheathing the scape below with their petioles. Berries bright red.

2. **Calla.** Flowers (at least the lower ones) *perfect*, covering the whole spadix. Spathe open and spreading, with a white upper surface, tipped with an abrupt point. Scape from a creeping rootstock. Leaves not net-veined, simple, heart-shaped.

** *Leaves not linear. Flowers with a perianth of 4 sepals. Spadix surrounded by a spathe.*

3. **Symplocar'pus.** Leaves all radical, very large and veiny, appearing after the spathes, which are close to the ground and are produced very early in spring. Flowers perfect, their ovaries immersed in the spadix, the latter globular and surrounded by the shell-shaped spathe. Sepals hooded. Stamens 4. Fruit consisting of the soft enlarged spadix in which the seeds are sunk.

*** *Leaves linear, sword-shaped. Spadix on the side of the scape. Flowers with a perianth of 6 sepals. No spathe.*

4. **Ac'orus.** Scape 2 edged, resembling the leaves, the cylindrical spadix borne on one edge. Sepals hollowed. Stamens 6.

1. ARISÆ'MA, Martins. INDIAN TURNIP.

1. **A. triphyl'lum**, Torr. (INDIAN TURNIP.) For full description and engraving of this plant see Part I., sections 94-97.

2. **A. Dracontium**, Schott., (GREEN DRAGON) is reported from low grounds near London, Ont. Leaf usually solitary, pedately divided into 7-11 oblong-lanceolate pointed leaflets. Spathe convolute, pointed; the slender point of the spadix extending beyond it.

2. CALLA, L. WATER ARUM.

C. palustris, L. (MARSH CALLA.) This plant is fully described and illustrated in Part I., section 98.

3. SYMPLOCAR'PUS, Salisb. SKUNK CABBAGE.

S. fœ'tidus, Salisb. Leaves 1-2 feet long, ovate or heart-shaped, short-petioled. Spathe purplish and yellowish, incurved. Plant with skunk-like odour.—Bogs and wet places; not common northward.

4. AC'ORUS, L. SWEET FLAG. CALAMUS.

A. Cal'amus, L. Scape much prolonged beyond the spadix. —Swamps and wet places.

ORDER XCI. **LEMNA'CEÆ.** (DUCKWEED FAMILY.)

Very small plants floating about freely on the surface of ponds and ditches, consisting merely of a little frond with a single root or a tuft of roots from the lower surface, and producing minute monœcious flowers from a cleft in the edge of the frond. The flowers are rarely to be seen. The commonest representative with us is

Lemna polyrrhi'za, consisting of little roundish green fronds (purplish beneath) about ¼ of an inch across, and with a *cluster* of little roots from the under surface.

L. minor is also found. *Root single.*

ORDER XCII. **TYPHA'CEÆ.** (CAT-TAIL FAMILY.)

Aquatic or marsh herbs with linear sword-shaped leaves, erect or floating, and monœcious flowers, either in separate heads or on different parts of the same spike or spadix, but without a spathe and destitute of true floral envelopes. Fruit an achene, 1-seeded.

1. **Ty'pha.** Flowers in a very dense and long cylindrical terminal spike, the upper ones staminate, the lower pistillate, the ovaries long-stalked and surrounded by copious bristles forming the down of the fruit. Leaves sword-shaped, erect, sheathing the stem below.
2. **Sparga'nium.** Flowers in separate globular heads along the upper part of the stem, the higher ones staminate, the lower pistillate, each ovary sessile and surrounded by a few scales not unlike a calyx. Both kinds of heads leafy-bracted. Leaves flat or triangular, sheathing the stem with their bases.

1. TYPHA, Tourn. CAT-TAIL FLAG.

1. **T. latifo'lia,** L. (COMMON CAT-TAIL.) Stem 5-8 feet high. Leaves *flat*. No space between the staminate and pistillate parts of the spike.—Marshy places.

2. **T. angustifo'lia,** L. (NARROW-LEAVED or SMALL CAT-TAIL.) Leaves *channelled* toward the base, narrowly linear. The two parts of the spike usually with an interval between them.— Central and eastern Ontario.

2. SPARGA'NIUM, Tourn. BUR-REED.

1. **S. eurycar'pum,** Engelm. Stem erect, stout, 2-4 feet high. Leaves mostly flat on the upper side, keeled and hollow-sided on the lower. Heads several, panicled-spiked, the pistillate an inch across in fruit. Nutlets or achenes with a broad abruptly-pointed top.—Borders of slow waters and ponds.

2. **S. simplex,** Hudson, var. **angustifo'lium,** Gray. (*S. affine*, Schnitzlein, in Macoun's Catalogue.) Stem slender, erect, 1-2 feet high; *the leaves usually floating*, long and narrowly linear. Heads several, usually in a simple row, the pistillate *supra-axillary*, about half an inch across. Nutlets pointed at both ends.

Var. **Nuttallii,** Engel., (*S. simplex*, Hudson, in Macoun's Catalogue,) has the lower *pistillate heads axillary*, and the fruiting heads perhaps a little larger.—In slow streams.

ORDER XCIII. **NAIADA'CEÆ.** (PONDWEED FAMILY.)

Immersed aquatic herbs, with jointed stems and sheathing stipules. Leaves immersed or floating. Flowers (in our common genus) perfect, in spikes or clusters, with 4 sepals, 4 stamens, and 4 ovaries; the spikes generally raised on peduncles to the top of the water. Plants of no very great interest. The most obvious characters of a few species are given here.

POTAMOGE'TON, Tourn. PONDWEED.

1. **P. natans,** L. *Submersed leaves grass-like or capillary;* upper stipules *very long, acute.* Spikes cylindrical, all out of the water. Stem hardly branched. Floating leaves *long-petioled,* elliptical, with a somewhat heart-shaped base, with a blunt apex, 21-29-nerved.

2. **P. Claytonii,** Tuckerman. *Stem compressed.* Submersed leaves *linear,* 2-5 inches long, 2-ranked, 5-nerved; stipules *obtuse.* Floating leaves *short-petioled,* chiefly opposite, oblong, 11-17-nerved. Spikes all above water.

3. **P. amplifo'lius,** Tuckerman. Submersed leaves large, *lanceolate or oval,* acute at each end, recurved, wavy; *stipules long and tapering.* Floating leaves large, oblong or lance-ovate, or slightly cordate, long-petioled, 30-50-nerved.

4. **P. gramin'eus,** L. Submersed leaves lanceolate or linear-lanceolate, upper ones petioled, lower ones sessile. *Stipules obtuse.* Floating leaves with a short blunt point, 9-15-nerved. Var. **heterophyl'lus,** Fries., (the common form) has the lower leaves shorter, lanceolate, and more rigid.

5. **P. lucens,** L., var. **minor,** Nolte. *Leaves all submersed,* more or less petioled, oval or lanceolate, *mucronate, shining.* Stem branching.

6. **P. perfolia'tus,** L. Leaves all submersed, varying in width from *orbicular to lanceolate, clasping by a heart-shaped base.* Stem branching.

7. **P. compressus,** Fries. (*P. zosterœfolius,* Schum., in Macoun's Catalogue.) Leaves all submersed, linear, grass-like, sessile, with three large nerves and many fine ones. Stem branching, *wing-flattened.* Stipules free from the sheathing base of the leaf.

8. **P. pectina'tus,** L. Leaves all submersed, bristle-shaped. Stem repeatedly forking, filiform. *Spikes interrupted, on long slender peduncles.* Stipules united with the sheathing base of the leaf.

ALISMACEÆ. 147

II. PETALOI'DEOUS DIVISION.

Flowers with a perianth coloured like a corolla.

ORDER XCIV. **ALISMA'CEÆ.** (WATER-PLANTAIN F.)

Marsh herbs, with flowers having 3 distinct sepals and 3 distinct petals, pistils either apocarpous or separating at maturity into distinct carpels, and hypogynous stamens 6-many. Flowers on scapes or scape-like stems. Leaves sheathing at the base, either rush-like or, when broad, mostly heart-shaped or arrow-shaped.

Synopsis of the Genera.

* *Calyx and corolla both greenish. Carpels more or less united, but spreading at maturity. Leaves rush-like and fleshy, or grass-like.*

1. **Triglo'chin.** Flowers small, in a spike or close raceme, without bracts. Carpels *united to the top;* when ripe, splitting away from a central persistent axis.

2. **Scheuchze'ria.** A low bog-herb, with a creeping jointed rootstock, and grass-like leaves. Stamens 6. Carpels 3, globular, *nearly distinct.*

(These two genera are included in Naiadaceæ in Macoun's Catalogue.)

** *Calyx green, persistent. Corolla white. Pistil apocarpous. Leaves with distinct blades and petioles.*

3. **Alis'ma.** *Flowers perfect.* Stamens usually 6. Carpels numerous, in a ring. Leaves all radical. Scape *with whorled panicled branches.*

4. **Sagitta'ria.** *Flowers monœcious, sometimes diœcious.* Stamens numerous. Carpels numerous, in more or less globular heads. Leaves arrow-shaped, but varying greatly. Flowers mostly in whorls of 3 on the scapes, the sterile ones uppermost.

1. TRIGLO'CHIN, L. ARROW-GRASS.

1. **T. palus'tre,** L. A slender rush-like plant, 6-18 inches high, found growing in bogs northward. *Carpels 3,* awl-pointed at the base, splitting away from below upwards. Spike or raceme slender, 3 or 4 inches long.

2. **T. marit'imum,** L., is also found occasionally. The whole plant is stouter than No. 1, and the carpels are usually 6 in number.

2. SCHEUCHZE'RIA, L. SCHEUCHZERIA.

S. palustris, L. Stem zigzag. Flowers in a loose terminal raceme, with sheathing bracts.—Bogs.

3. ALIS'MA, L. WATER-PLANTAIN.

A. planta'go, L., var. **America'num,** Gray. Leaves long-petioled, mostly oblong-heart-shaped, but often narrower, 3-9-nerved or ribbed, and with cross veinlets between the ribs. Flowers small, white, in a large and loose compound panicle.—Low and marshy places, often growing in the water.

4. SAGITTA'RIA, L. ARROW-HEAD.
* *Filaments narrow, as long as the anthers.*

1. **S. varia'bilis,** Engelm. Very variable in size and in the shape of the leaves. Scape angled.—Common everywhere in shallow water.

* * *Filaments very short, with enlarged mostly glandular base.*

2. **S. heterophyl'la,** Pursh. Scape weak and at length procumbent. Leaves lanceolate or lance-ovate, entire, or with one or two narrow basal sagittate appendages.

3. **S. gramin'ea,** Michx. Scape very slender, erect. Leaves varying from ovate-lanceolate to linear, *scarcely ever sagittate*.

ORDER XCV. HYDROCHARIDA'CEÆ. (FROG'S-BIT F.)

Aquatic herbs, with diœcious or polygamo-diœcious flowers on scape-like peduncles from a kind of spathe of one or two leaves, the perianth in the fertile flowers of 6 pieces united below into a tube which is adherent to the ovary. Stigmas 3. Fruit ripening under water.

Synopsis of the Genera.

1. **Anach'aris.** Growing under water, the pistillate flowers alone coming to the surface. Stem leafy and branching. Perianth of the fertile flowers with a 6-lobed spreading limb, the tube prolonged to an extraordinary length, thread-like. Leaves crowded, pellucid, 1-nerved, sessile, whorled in threes or fours. Stamens 3-9.

2. **Vallisne'ria.** Nothing but the pistillate flowers above the surface, these on scapes of great length, and after fertilization drawn below the surface by the spiral coiling of the scapes. Tube of the perianth not prolonged. Leaves linear, thin, long and ribbon-like.

(In both genera the staminate flowers break off spontaneously and float on the surface around the pistillate ones, shedding their pollen upon them.)

1. ANACH'ARIS, Richard. WATER-WEED.

A. Canadensis, Richard. (*Elodéa Canadensis*, Planchon, in Macoun's Catalogue.)—Common in slow waters.

2. **VALLISNE'RIA**, Micheli. TAPE-GRASS. EEL-GRASS.

V. spira'lis, L. Leaves 1-2 feet long.—Common in slow waters.

ORDER XCVI. **ORCHIDA'CEÆ.** (ORCHIS FAMILY.)

Herbs, well marked by the peculiar arrangement of the stamens, these being *gynandrous*, that is, borne on or adherent to the stigma or style. There is also usually but a single stamen, of two rather widely separated anthers, but in the last genus of the following list there are 2 distinct stamens, with the rudiment of a third at the back of the stigma. As explained in Part I., sections 90-93, the Orchids as a rule require the aid of insects to convey the *pollinia*, or pollen-masses, to the stigma, but occasionally it happens that when the anther-cells burst open the pollinia fall forward and dangle in front of the viscid stigma beneath, being sooner or later driven against it either by the wind or by the head of some insect in pursuit of honey. In all cases where the student meets with an Orchid in flower, he should, by experiment, endeavour to make himself acquainted with the method of its fertilization.

The Orchis Family is a very large one, there being probably as many as 3,000 different species, but the greater number are natives of tropical regions. Many of them are *epiphytes*, or air-plants, deriving their support chiefly from the moisture of the air, through their long aerial roots which never reach the ground. The perianth in many species, and particularly the *labellum*, or lip, assumes the most fantastic shapes, making the plants great favourites for hot-house cultivation. In Canada, the representatives of this great Order, though not very numerous, are among the most interesting and beautiful of our wild flowers. They are, as a rule, bog-plants, and will be found in flower in early summer.

Synopsis of the Genera.

* *Anther only one, but of 2 cells, these separated in the first genus.*
+ *Lip with a spur underneath. Anther on the face of the stigma.*

1. **Orchis.** The 3 sepals and 2 of the petals erect and arching over the centre of the flower; the lip turned down. The 2 glands or viscid disks at the base of the pollen-masses enclosed in a little pouch just over the concave stigma. Leaves 2, large. Flowers few, in a spike.

2. **Habena'ria.** The lateral sepals usually spreading. The glands or viscid disks of the pollen-masses not enclosed in a covering. Flowers in spikes.

+ + *Lip without a spur. Anther on the back of the column. Flowers small, white, in a slender spike.*

3. **Spiran'thes.** Spike (of white or whitish flowers) more or less spirally twisted. Sepals and petals narrow and generally connivent. Lip oblong; the lower part embracing the column, *and with a protuberance on each side at the base.*

4. **Goodye'ra.** Flowers very much as in Spiranthes, *but the lip sac-shaped, and without protuberances at the base.* Leaves white-veiny, in a tuft at the base of the scape.

+ + + *Lip without a spur. Anther on the apex of the style, hinged like a lid.*
++ *Pollen-masses 2 or 4, powdery or pulpy, without stalk or gland.*

5. **Lis'tera.** Flowers small, greenish or brownish-purple, in a spike or raceme. Stem bearing a pair of opposite sessile roundish leaves near the middle. Lip flat, *mostly drooping, 2-lobed at the apex.*

6. **Calopo'gon.** Ovary not twisted, the lip consequently turned towards the stem. Flowers large, pink-purple, 2–6 at the summit of the scape; the lip spreading at the outer end and beautifully bearded above with coloured hairs. Leaf grass-like, only one. Pollen-masses 4.

7. **Arethu'sa.** Flower solitary, large, rose-purple. Lip dilated, recurved-spreading at the end. Sepals and petals lanceolate, nearly alike, arching over the column. Pollen-masses 4. Scape low, sheathed, from a globular solid bulb, with a single linear nerved leaf hidden in the sheaths of the scape.

8. **Pogo'nia.** Flower solitary, irregular, large, sweet-scented, pale rose-colour or white. Column club-shaped. Lip crested and fringed. Pollen-masses 2. Stem 6–9 inches high, with a single oval or lance-oblong leaf near the middle, and a smaller one, or bract, near the flower.

++ ++ *Pollen-masses 4, smooth and waxy, attached directly to a large gland: no stalks.*

9. **Calyp'so.** Flower solitary, large, showy, variegated with purple, pink, and yellow. Lip large, inflated, sac-shaped, 2 pointed under the apex. Scape short, from a solid bulb, with a single ovate or slightly heart-shaped leaf below.

++ ++ ++ *Pollen-masses 4: no stalks or glands.*

10. **Micros'tylis.** Small herbs from solid bulbs; the scape bearing a single leaf and a raceme of minute greenish flowers. Column very small, terete, with 2 teeth at the top, and the anther between them. Petals thread-like or linear, spreading.

11. **Lip'aris.** Small herbs, from solid bulbs; the low scape bearing 2 radical leaves and a raceme of a few greenish flowers. Column elongated incurved, margined at the apex. Petals thread-like or linear, spreading. Anther lid-like.

12. **Corallorhi'za.** Brownish or yellowish plants, with the small dull flowers in spikes on scapes which are leafless or have mere sheaths instead of leaves. Rootstocks branching and coral-like. *Perianth gibbous or slightly spurred below.* Lip with 2 ridges on the inner part of the face.

13. **Aplec'trum.** Somewhat like the last, but the perianth is *not gibbous below*, and the rootstock, instead of being coral-like, is slender, and produces each year a solid bulb or corm. Lip with 3 ridges on the palate. Scape with 3 greenish sheaths below.

* * *Anthers 2, one on each side of the stigma, and a triangular body, which is the rudiment of a third, at the back of the stigma. Pollen loose and powdery or pulpy.*

14. **Cypripe'dium.** Lip *a large inflated sac*, into the mouth of which the style is declined. Sepals and the other petals much alike, the former apparently only 2, two of them being generally united into one under the lip. Leaves large, many-nerved. Flowers solitary or few.

1. ORCHIS, L. Orchis.

O. specta'bilis, L. (SHOWY ORCHIS.) Scape 4-angled, 4-7 inches high, bearing a few flowers in a spike. The arching upper lip pink-purple, the *labellum* white; each flower in the axil of a leaf-like bract.—Rich woods.

2. HABENA'RIA, Willd., R. Br. REIN-ORCHIS.

1. **H. tridenta'ta,** Hook. *Spike few-flowered*, the flowers very small, greenish-white. *Lip wedge-shaped, truncate and 3-toothed at the apex. Spur slender, longer than the ovary*, curved upwards. Stem less than a foot high, slender, with one oblanceolate leaf below and 2 or 3 much smaller ones above.—Wet woods.

2. **H. vires'cens,** Spreng. Stem 10-20 inches high. Spike of small greenish flowers at first dense, with the bracts longer than the flowers, at length long and loose. Lip oblong, almost truncate at the tip; a tooth on each side at the base, and a nasal protuberance on the face. Spur slender, club-shaped. Leaves ovate-oblong or oblong-lanceolate, the upper ones gradually narrowing and passing into bracts.—Wet places.

3. **H. vir'idis,** R. Br., var. **bracteata,** Reichenbach. (*H. bracteata*, R. Br., in Macoun's Catalogue.) *Spike many-flowered*, close. Flowers small, *greenish*. Lip oblong-linear, 2-3-lobed at the tip, *much longer than the very short and sac-like spur*. Stem 6-12 inches high, *leafy*, the lower leaves obovate, the upper oblong or lanceolate, gradually reduced to bracts much longer than the flowers.

4. **H. hyperbo′rea,** R. Br. Spike many-flowered, *long and dense.* Flowers small, greenish. Lip lanceolate, *entire, about the same length as the slender incurved spur.* Stem 6-24 inches high, *very leafy, the leaves lanceolate and erect,* and the bracts longer than the flowers.—Bogs and wet woods.

5. **H. dilata′ta,** Gray. Not unlike No. 4, but more slender and with *linear leaves* and *white flowers.*

6. **H. rotundifo′lia,** Richardson. (*Orchis rotundifolia,* Gray, in Macoun's Catalogue.) Spike few-flowered, loose. Flowers rose-purple, *the lip usually white, spotted with purple, 3-lobed,* the middle lobe larger and notched, longer than the slender spur. Stem 5-9 inches high, *naked and scape-like above, bearing a single roundish leaf* at the base.—Bogs and wet woods.

7. **H. obtusa′ta,** Richardson. Stem as in the last, but the leaf is obovate or spathulate-oblong. Spike few-flowered, the flowers *greenish-white.* Upper sepal broad and rounded, the others and the petals lance-oblong. Lip *entire,* deflexed, as long as the tapering and curving spur.—Bogs.

8. **H. Hook′eri,** Torr. Spike many-flowered, *strict. Flowers yellowish-green,* the lip lanceolate, pointed, incurved; petals lance-awl-shaped. *Spur slender, acute, nearly an inch long.* Stem scape-like above, *2-leaved at the base, the leaves orbicular.*—Woods.

9. **H. orbicula′ta,** Torr. Spike many-flowered, *loose and spreading.* Flowers *greenish-white.* Lip narrowly linear, obtuse. *Spur curved, more than an inch long, thickened towards the apex.* Scape 2-leaved at the base, the leaves *very large, orbicular, and lying flat on the ground,* shining above, silvery beneath.—Rich woods.

10. **H. blephariglot′tis,** Hook. (WHITE FRINGED-ORCHIS.) Spike many-flowered, open. *Flowers white,* very handsome; *the lip fringed, but not lobed,* at the apex. *Spur thread-shaped, three times as long as the lip.* Stem a foot high, leafy; the leaves oblong or lanceolate, the bracts shorter than the ovaries.—Peat-bogs, &c.

11. **H. leucophæ′a,** Gray. (GREENISH FRINGED-ORCHIS.) Spike as in the last, but the flowers *greenish or yellowish-white.* Petals obovate, minutely cut-toothed. *Lip 3-parted above the stalk-like base, the divisions fan-shaped, fringed.* Spur *gradually*

thickened downward, longer than the ovary. Stem leafy, 2-4 feet high. Leaves oblong-lanceolate; bracts a little shorter than the flowers.—Wet meadows.

12. **H. la'cera**, R. Br. (RAGGED FRINGED-ORCHIS.) Like the last, but the *petals are oblong-linear and entire*. The divisions of the lip also are *narrow* and the *fringe is less copious*.—Bogs and rich woods.

13. **H. psyco'des**, Gray. (PURPLE FRINGED-ORCHIS.) Spike cylindrical, many-flowered, the *flowers pink-purple*, fragrant. Lip fan-shaped, 3-parted above the stalk-like base, *the divisions fringed*. Spur curved, somewhat thickened downward, very long. —Low grounds.

3. SPIRAN'THES, Richard. LADIES' TRESSES.

1. **S. Romanzovia'na**, Chamisso. Spike dense, oblong or cylindrical. Flowers *pure white, in 3 ranks in the spike.* Lip ovate-oblong, contracted below the wavy recurved apex. Stem 5-15 inches high, leafy below, leafy-bracted above; the leaves oblong-lanceolate or linear.—Cool bogs.

2. **S. gra'cilis**, Bigelow. *Flowers in a single spirally twisted rank* at the summit of the very slender scape. Leaves with blades all in a cluster at the base, ovate or oblong. Scape 8-18 inches high.—Sandy plains and pine barrens.

4. GOODYE'RA, R. Br. RATTLE-SNAKE PLANTAIN.

1. **G. repens**, R. Br. *Flowers in a loose 1-sided spike.* Lip with a recurved tip. Scape 5-8 inches high. Leaves thickish, petioled, intersected with whitish veins.—Woods, usually under evergreens.

2. **G. pubes'cens**, R. Br. *Spike not 1-sided.* Plant rather larger than the last, and the leaves more strongly white-veined. —Rich woods.

5. LIS'TERA, R. Br. TWAYBLADE.

1. **L. corda'ta**, R. Br. Raceme crowded; pedicels not longer than the ovary. Lip linear, 2-cleft. Column very short.—Damp cold woods.

2. **L. convallarioi'des**, Nutt. Raceme loose and slender; pedicels longer than the ovary. Lip wedge-oblong, 2-lobed. Column longer than in the last.—Damp thickets.

6. **CALOPO'GON**, R. Br. CALOPOGON.

C. pulchel'lus, R. Br. Leaf linear. Scape a foot high. Flowers an inch across.--Bogs.

7. **ARETHU'SA**, Gronov. ARETHUSA.

A. bulbo'sa, L. A beautiful little bog-plant, bearing a single large flower (rarely 2), with the lip bearded-crested on the face.

8. **POGO'NIA**, Juss. POGONIA.

P. ophioglossoi'des, Nutt. A bog-plant. Sepals and petals nearly equal and alike. Root of thick fibres.

9. **CALYP'SO**, Salisb. CALYPSO.

C. borea'lis, Salisb. A beautiful little plant growing in mossy bogs. The lip woolly inside; the petals and sepals resembling each other, lanceolate, sharp-pointed. Column winged.

10. **MICROS'TYLIS**, Nutt. ADDER'S-MOUTH.

1. **M. monophyl'los**, Lindl. Leaf sheathing the base of the stem, ovate-elliptical. Raceme spiked, long and slender. Lip long-pointed.—Cold swamps.

2. **M. ophioglossoi'des**, Nutt. Leaf near the middle of the stem, ovate, clasping. Raceme *short*. Lip 3-toothed.—Damp woods, not so common as the last.

11. **LIP'ARIS**, Richard. TWAYBLADE.

L. Loese'lii, Richard. Lip yellowish-green, mucronate, shorter than the unequal petals and sepals. Leaves elliptical-lanceolate or oblong, keeled.—Bogs.

12. **CORALLORHI'ZA**, Haller. CORAL-ROOT.

1. **C. inna'ta**, R. Br. Flowers small; the lip whitish or purplish, often crimson-spotted, 3-lobed above the base. Spur very small. Stem slender, brownish-yellow, *with a few-flowered spike*.—Swamps.

2. **C. multiflo'ra**, Nutt. *Spike many-flowered*. Stem purplish, stout. Lip deeply 3-lobed. Spur more prominent than in No. 1. —Dry woods.

3. **C. Macræ'i**, Gray. (*C. striata*, Lindl., in Macoun's Catalogue.) Spike crowded, of numerous large flowers, *all the parts*

of the perianth strikingly marked with 3 dark lines. Lip not lobed. Spur none, but the base of the perianth gibbous.—Rich woods: not common.

13. **APLEC'TRUM**, Nutt. PUTTY-ROOT. ADAM-AND-EVE.

A. hyema'le, Nutt. Scape a foot high. Perianth greenish-brown.—Rich mould in woods.

14. **CYPRIPE'DIUM**, L. LADY'S SLIPPER. MOCCASIN-FLOWER.

* *The three sepals separate.*

1. **C. arieti'num**, R. Br. (RAM'S-HEAD LADY'S SLIPPER.) The smallest species. Stem slender, 6-10 inches high, leafy. Leaves 3 or 4, elliptical-lanceolate, nearly smooth. Lip only half an inch long, red and whitish veiny, prolonged at the apex into a deflexed point.—Swamps; rare.

* * *Two sepals united into one piece under the lip.*

2. **C. parviflo'rum**, Salisb. (SMALLER YELLOW LADY'S SLIPPER.) Stem leafy to the top, 1-3-flowered. Lip yellow, *flattish above*, rather less than an inch long. Sepals and petals wavy-twisted, brownish, pointed, longer than the lip.—Bogs and wet woods.

3. **C. pubes'cens**, Willd. (LARGER YELLOW L.) Lip flattened *laterally*, rounded above, larger than in No. 1, but the two species are not sufficiently distinct.

4. **C. specta'bile**, Swartz. (SHOWY L.) *Lip very large, white, pinkish in front*. Sepals and petals *rounded, white*, not longer than the lip.—Bogs.

5. **C. acau'le**, Ait. (STEMLESS L.) *Scape naked*, 2-leaved at the base, 1-flowered. *Lip rose-purple*, split down the whole length in front, veiny. Sepals and petals greenish.—Dry or moist woods, under evergreens.

ORDER XCVII. **IRIDA'CEÆ**. (IRIS FAMILY.)

Herbs with equitant leaves and perfect flowers. The 6 petal-like divisions of the perianth in 2 (similar or dissimilar) sets of 3 each; the tube adherent to the 3-celled ovary. Stamens 3, distinct or monadelphous, opposite the 3 stigmas, and with anthers extrorse, that is, on the outside of the filaments, facing the divisions of the perianth and opening on that side. Flowers from leafy bracts. (See Part I., sections 88 and 89.)

Synopsis of the Genera.

1. **Iris.** The 3 outer divisions of the perianth reflexed, the 3 inner erect and smaller. Stamens distinct, the anther of each concealed under a flat and petal-like arching stigma. The styles below adherent to the tube of the perianth. Pod 3-angled. Flowers blue, large and showy. Leaves sword-shaped or grass-like.
2. **Sisyrin'chium.** The 6 divisions of the perianth alike, spreading. Stamens monadelphous. Stigmas thread-like. Pod globular, 3-angled. Stems 2-edged. Leaves grass-like. Flowers blue, clustered, from 2 leafy bracts. Plants low and slender.

1. IRIS, L. FLOWER-DE-LUCE.

1. **I. versic'olor,** L. (LARGER BLUE FLAG.) Stem stout and leafy, from a thickened rootstock. Leaves sword-shaped. Flowers violet-blue, 2 or 3 inches long. Inner petals much smaller than the outer.—Wet places.

2. **I. lacus'tris,** Nutt. (LAKE DWARF IRIS.) Stem low, 3–6 inches high. Inner petals nearly equal to the outer. Tube of the perianth slender, less than an inch long, dilated upwards, rather shorter than the divisions of the perianth. Leaves lanceolate, 3–5 inches long.—Shore of Lake Huron.

2. SISYRIN'CHIUM, L. BLUE-EYED GRASS.

S. Bermudia'na, L., var. **anceps,** Gray. (*S. anceps,* Cav., in Macoun's Catalogue.) A pretty little plant, rather common in moist meadows among grass. The divisions of the delicate blue perianth obovate, notched at the end, and bristle-pointed from the notch. Roots fibrous.

ORDER XCVIII. AMARYLLIDA'CEÆ. (AMARYLLIS F.)

Bulbous and scape-bearing herbs, with linear flat root-leaves, and regular and perfect 6-androus flowers, the tube of the petal-like 6-parted perianth adherent to the 3-celled ovary. Lobes of the perianth imbricated in the bud. Style single. Anthers introrse. Represented with us by one species of the genus

HYPOX'YS, L. Star-grass.

H. erecta, L. A small herb sending up a slender scape from a solid bulb. Leaves linear, grass-like, longer than the umbellately 1–4-flowered scape. Perianth hairy and greenish outside, yellowish within, 6-parted nearly down to the ovary. Stamens 6, sagittate. Pod indehiscent, crowned with the withered perianth.—Meadows and open woods.

Order XCIX. **DIOSCOREA'CEÆ.** (Yam Family.)

Represented with us by the genus

DIOSCORE'A, Plumier. Yam.

D. villo'sa, L. (Wild Yam-root.) A slender twiner with knotted rootstocks, and net-veined, heart-shaped, 9–11-ribbed, petioled leaves. Flowers diœcious, small, in axillary racemes. Stamens 6. Pod with three large wings.—Reported only from the warm and sheltered valley lying between Hamilton and Dundas, Ont., and the banks of the Thames at London, Ont.

Order C. **SMILA'CEÆ.** (Smilax Family.)

Climbing plants, more or less shrubby, with alternate *ribbed and net-veined* petioled leaves, and small diœcious flowers in umbels. Perianth regular, of 6 greenish sepals, free from the ovary. Stamens as many as the sepals, with 1-celled anthers. Ovary 3-celled, surmounted by 3 sessile spreading stigmas. Fruit a small berry. Represented by the single genus

SMILAX, Tourn. Green-Brier. Cat-Brier.

(*Included in Liliaceæ, in Macoun's Catalogue.*)

1. **S. his'pida,** Muhl. *Stem woody, densely covered below with long weak prickles.* Leaves large, ovate or heart-shaped, pointed, thin, 5–9-nerved. Peduncles of the axillary umbels *much longer than the petioles.* Berry black.—Moist thickets.

2. **S. rotundifo'lia,** L., var. **quadrangularis,** Gray. (*S. quadrangularis,* Pursh, in Macoun's Catalogue.) *Stem woody, it and the branches armed with scattered prickles. Branches 4-angular.* Peduncles *not longer than the petioles.* Leaves ovate, broader than long, slightly cordate. Berry blue-black.—Southwestern Ontario.

3. **S. herba'cea,** L. (CARRION-FLOWER.) *Stem herbaceous, not prickly.* Leaves ovate-oblong and heart-shaped, 7-9-ribbed, long-petioled, mucronate. Flowers carrion-scented. Berry bluish-black.—Meadows and river-banks.

ORDER CI. **LILIA'CEÆ.** (LILY FAMILY.)

Herbs, distinguished as a whole by their regular and symmetrical flowers, having a 6-leaved perianth (but 4-leaved in one species of Smilacina) free from the usually 3-celled ovary, and as many stamens as divisions of the perianth (*one before each*) with 2-celled anthers. Fruit a pod or berry, generally 3-celled. The outer and inner divisions of the perianth coloured alike, except in the genus Trillium. (See Part I., sections 82-87, for description of typical plants of this Order.)

Synopsis of the Genera.

* *Leaves net-veined, all in one or two whorls. The stem otherwise naked, rising from a fleshy rootstock. Styles 3.*

1. **Tril'lium.** Leaves 3, in a whorl at the top of the stem. Divisions of the perianth in 2 sets, the outer green, the inner coloured. (See Part I., sections 85 and 86.)

2. **Mede'ola.** Leaves in 2 whorls, the lower near the middle of the stem, and consisting of 5-9 leaves, the upper of (generally) 3 small leaves, near the summit. Stem tall, covered with loose wool. Flowers small, in an umbel. Divisions of the perianth alike, greenish-yellow, recurved. Anthers turned outwards. Styles thread-shaped. Berry globular or nearly so, dark purple.

* * *Leaves straight-veined, linear, grass-like, alternate. Stems simple or tufted. Styles 3.*

3. **Zygade'nus.** Flowers rather large, perfect or polygamous, greenish-white, in a few-flowered panicle ; the divisions of the perianth each with a conspicuous obcordate spot or gland on the inside, near the narrowing base. Stem smooth and glaucous, *from a coated bulb.*

4. **Tofield'ia.** Flowers small, perfect, greenish-white, in a terminal raceme or spike, which, however, develops from above downward; the pedicels in clusters of 3, from little involucres of 3 bracts. Pod triangular. Roots *fibrous*. Stem leafy at the base only, in our species *sticky*. Leaves 2-ranked, equitant.

* * * *Leaves straight-veined, but broad (not grass-like), alternate. Stem from a rootstock or fibrous roots, at all events not from a bulb. Style one at the base, but more or less divided into 3 above.*

+ *Perianth of completely separate pieces (polyphyllous).*

5. **Uvula'ria.** Stem leafy, *forking above.* Flowers yellow, at least an inch long, drooping, lily-like, usually solitary (but occasionally in pairs) at

LILIACEÆ.

the end or in the forks of the stem. Style *deeply* 3-cleft. Pod triangular. *Leaves clasping-perfoliate or sessile.*

6. **Clinto'nia.** *Stemless,* the naked scape sheathed at the base by 2, 3, or 4 large oblong or oval ciliate leaves. Flowers few, greenish-yellow, in an umbel at the top of the scape. Filaments long and slender. Style long, the stigmas hardly separate. Berry blue.

7. **Prosar'tes.** Downy low herbs, branching above. Flowers greenish, bell-shaped, rather large, solitary or in pairs, drooping on terminal slender peduncles. Sepals taper-pointed. Stigmas 3. Leaves ovate-oblong, taper-pointed, closely sessile, downy underneath. Berry oblong, pointed, red.

8. **Strep'topus.** Stem leafy and forking. Flowers small, not quite in the axils of the ovate clasping leaves, *on slender peduncles which are abruptly bent near the middle.* Anthers arrow-shaped, *2-horned at the apex.*

++ *Perianth of one piece (gamophyllous).*

9. **Smilaci'na.** Flowers small, white, in a terminal raceme. Perianth 6-parted, *but 4-parted in one species,* spreading. Style short and thick. Stigma obscurely lobed. Filaments slender.

10. **Polygona'tum.** Flowers small, greenish, nodding, *mostly in pairs* in the axils of the nearly sessile leaves. Perianth cylindrical, 6-lobed at the summit, *the 6 stamens inserted* on or above *the middle of the tube.* Stem simple, from a long and knotted rootstock. Leaves glaucous beneath.

* * * * *Leaves straight-veined, not grass-like. Stem from a coated or scaly bulb. Style 1, not divided above, but the stigma sometimes 3-lobed. Fruit a pod, splitting open midway between the partitions (loculicidal).*

11. **Lil'ium.** Stem leafy, from a scaly bulb, *the leaves often whorled or crowded.* Anthers at first erect, *at length versatile.* Style long, rather club-shaped. Stigma 3-lobed. Pod oblong. Flowers large and showy, one or more.

12. **Erythro'nium.** For full description see Part I., sections 82 and 83. (Dog's-tooth Violet.)

13. **Allium.** *Scape naked,* from a *coated* bulb. Flowers in an umbel, from a spathe. Style thread-like. *Strong-scented plants.*

1. TRIL'LIUM, L. WAKE-ROBIN.

1. **T. grandiflo'rum,** Salisb. (LARGE WHITE TRILLIUM.) *Leaves sessile,* longer than broad. *Peduncle erect.* Petals *white* (rose-coloured when old), *obovate.*—Rich woods.

2. **T. erectum,** L. (*T. erectum,* L., var. *atropurpureum,* Hook., in Macoun's Catalogue.) (PURPLE TRILLIUM.) *Leaves sessile,* about as broad as long. *Peduncle erect.* Petals dull

purple, *ovate.*—Rich woods. Var. **album**, *with greenish-white petals,* is found along with the purple form. It does not appear to be clearly distinguished from No. 1.

3. **T. cer'nuum**, L. *Leaves sessile* or nearly so, broadly rhomboid, abruptly pointed. *Peduncle recurved under the leaves.* Petals white, oblong-ovate, acute.—Chiefly eastward.

4. **T. erythrocar'pum**, Michx. (PAINTED TRILLIUM.) *Leaves distinctly petioled,* rounded at the base. Petals pointed, white, *with purple stripes inside at the base.*—Not uncommon northward in damp woods and low grounds.

2. MEDE'OLA, Gronov. INDIAN CUCUMBER-ROOT.

M. Virgin'ica, L. Stem 1-3 feet high.—Rich woods.

3. ZYGADE'NUS, Michx. ZYGADENE.

Z. glaucus, Nutt. Not uncommon in bogs and beaver-meadows northward. Leaves flat and pale.

4. TOFIELD'IA, Hudson. FALSE ASPHODEL.

T. glutino'sa, Willd. Stem and pedicels very sticky with dark glands. Leaves short.—Lake Huron coast.

5. UVULA'RIA, L. BELLWORT.

1. **U. grandiflo'ra**, Smith. Leaves clasping-perfoliate. Rootstock short.—Rich woods.

2. **U. sessilifo'lia**, L. Leaves *sessile* or partly clasping, lance-oblong. Rootstock *creeping.*—Eastward.

6. CLINTO'NIA, Raf. CLINTONIA.

C. borea'lis, Raf. Umbel 2-7-flowered. Leaves 5-8 inches long. Perianth pubescent outside.—Damp woods, often under evergreens.

7. PROSAR'TES, Don. PROSARTES.

P. lanugino'sa, Don. (*Disporum lanuginosa*, Don., in Macoun's Catalogue.)—Rich woods, western Ontario.

8. STREP'TOPUS, Michx. TWISTED-STALK.

S. ro'seus, Michx. Flowers rose-purple.—Damp woods.

LILIACEÆ.

9. SMILACI'NA, Desf. FALSE SOLOMON'S SEAL.

1. **S. racemo'sa,** Desf. (FALSE SPIKENARD.) *Raceme compound.* Stem pubescent, 2 feet high. Leaves many, oblong, taper-pointed, ciliate. Berries speckled with purple.—Rich woods and thickets.

2. **S. stella'ta,** Desf. *Raceme simple.* Stem nearly smooth, 1-2 feet high. Leaves 7-12, oblong-lanceolate, slightly clasping. Berries black.—Moist woods and copses.

3. **S. trifo'lia,** Desf. *Raceme simple.* Stem low (3-6 inches), glabrous. Leaves usually 3, oblong, the bases sheathing. Berries red.—Bogs.

4. **S. bifo'lia,** Ker., var. **Canadensis,** Gray. (*Maianthemum Canadense,* Desf., in Macoun's Catalogue.) Distinguished at once by the *4-parted perianth and the 4 stamens.* Raceme simple. Stem 3-5 inches high. Leaves usually 2, but sometimes 3.—Moist woods.

10. POLYGONA'TUM, Tourn. SOLOMON'S SEAL.

1. **P. biflo'rum,** Ell. (SMALLER SOLOMON'S SEAL.) Stem slender, 1-3 feet high. Leaves ovate-oblong or lance-oblong. Peduncles mostly 2-flowered. Filaments *hairy.*—Rich woods.

2. **P. gigante'um,** Dietrich, (GREAT S.) is occasionally met with westward and south-westward. The stem is taller and stouter than in the last, the peduncles *several-flowered,* and the filaments are *not hairy.*

11. LIL'IUM, L. LILY.

1. **L. Philadel'phicum,** L. (WILD ORANGE-RED LILY.) Divisions of the perianth *narrowed into claws* below, not recurved at the top. Flowers *erect,* 1-3, orange, spotted with purple inside. Leaves linear-lanceolate, the upper mostly in whorls of 5-8.—Sandy soil.

2. **L. Canadense,** L. (WILD YELLOW LILY.) Divisions of the perianth *recurred above the middle.* Flowers *nodding,* few, orange, spotted with brown inside. Leaves remotely whorled, 3-ribbed.—Swamps and wet meadows.

3. **L. super'bum,** L. (*L. Carolinianum,* Michx., in Macoun's Catalogue.) (TURK'S-CAP LILY.) Divisions of the perianth *very*

strongly recurved. Flowers nodding, *often numerous,* in a pyramidal raceme, bright orange, dark-purple-spotted within. Lower leaves whorled, 3-ribbed or nerved. Stem taller than either of the first two, 3-7 feet.—Rich low grounds, commoner southward and south-westward.

12. ERYTHRO'NIUM, L. Dog's-tooth Violet.

E. America'num, Smith. (Yellow Adder's Tongue.) Perianth light yellow, sometimes spotted at the base.—Copses and rich meadows.

13. ALLIUM, L. Onion. Leek.

1. **A. tricoc'cum,** Ait. (Wild Leek.) Leaves *flat, lance-oblong,* 5-9 inches long, 1-2 inches wide, appearing in early spring and withering before the flowers are developed. Sepals *white.* Pod strongly 3-lobed. Scape 9 inches high.—Rich woods.

2. **A. Canadense,** Kalm. (Wild Garlic.) Leaves *narrowly linear.* Ovary crested with 6 teeth. Umbel few-flowered, *often bearing a head of bulbs instead of flowers.* Sepals pale rose-colour.—Eastward, in moist meadows.

Order CII. JUNCA'CEÆ. (Rush Family.)

Grass-like or sedge-like plants, with, however, flowers similar in structure to those of the last Order. Perianth greenish and glumaceous, of 6 divisions in 2 sets of 3 each. Stamens 6 (occasionally 3). Style 1. Stigmas 3. Pod 3-celled, or 1-celled with 3 placentæ on the walls. The plants of the Order are not of any very great interest to the young student, and the determination of the species is rather difficult. A brief description of a few of the most common is given here, as an easy introduction to the study of the Order with the aid of more advanced text-books.

Synopsis of the Genera.

1. **Lu'zula.** Plant less than a foot high. Leaves linear or lance-linear, flat, *usually hairy.* Pod 1-celled, 3-seeded. Flowers in umbels or in spikes. Plants usually growing in *dry* ground.

2. **Juncus.** Plants *always smooth,* growing in water or *wet soil;* Flowers small, greenish or brownish, panicled or clustered. Pod 3-celled, *many-seeded.*

JUNCACEÆ.

1. LU'ZULA, DC. WOOD-RUSH.

1. **L. pilo'sa**, Willd. Flowers umbelled, long-peduncled, brown-coloured. Sepals pointed.—Shady banks.

2. **L. campestris**, DC., has the flowers (light brown) in 4-12 spikes, the spikes umbelled. Sepals bristle-pointed.—Fields and woods.

2. JUNCUS, L. RUSH.

Scapes simple and leafless, but with sheaths at the base. Flowers in sessile panicles, apparently from the side of the scape, owing to the involucral leaf being similar to and continuing the scape.

1. **J. effu'sus**, L. (COMMON or SOFT RUSH.) Scape 2-4 feet high, soft and pliant, furnished at the base with merely *leafless sheaths*, the inner sheaths awned. Panicle many-flowered. Flowers small, *greenish, only 1 on each pedicel*. Stamens 3. Pod *greenish-brown*, triangular-obovate, *not pointed*.—Marshes.

2. **J. filifor'mis**, L., has a very slender scape (1-2 feet high), fewer flowers than No. 1, and *6 stamens* in each. Pod *greenish*, broadly-ovate, and *short-pointed*. *No leaves*.

3. **J. Bal'ticus**, Dethard. Scape *rigid*, 2-3 feet high. *No leaves*. Panicle loose. Flowers brownish. Pod elliptical, somewhat triangular, obtuse but pointed, *deep-brown*.

* * *Stem leafy at the base or throughout; the leaves flat or channelled, but never knotted. Panicle terminal.*

4. **J. bufo'nius**, L. Stem *leafy*, slender, 3-9 inches high, branching from the base. Panicle terminal, spreading. Flowers greenish, single on the *pedicels*. Sepals awl-pointed, the outer set much longer than the inner and than the blunt pod. *Stamens 6*.—Ditches along roadsides.

5. **J. ten'uis**, Willd. Stems *leafy below*, wiry, 9-18 inches high, *simple*, tufted. Panicle loose, shorter than the slender involucral leaves. Flowers greenish, single on the pedicels; the sepals longer than the blunt pod. *Seeds white-pointed at both ends.*—Open low grounds.

* * * *Stem leafy; the leaves terete or laterally compressed, knotted by internal cross-partitions. Panicle terminal, the flowers mostly in heads.*

← *Stamens 6.*

6. **J. pelocar'pus**, E. Meyer. Stems slender and erect, 6-18 inches high. Leaves few, thread-like, slightly knotted. Flowers greenish with red, *single or in pairs* in the forks and along one

side of the branches of the panicle, and often with accompanying tufts of leaves. The 3 inner sepals longer than the outer ones, but shorter than the oblong taper-beaked 1-celled pod. Seeds obovate, short-pointed.

7. **J. alpi'nus**, Villars, var. **insignis**, Fries. Stems erect, 9-18 inches high, with 1 or 2 slender leaves. Branches of the meagre panicle erect, bearing *numerous* greenish or brownish *heads* of 3-6 flowers each. Outer sepals mucronate or cuspidate, and longer than the rounded inner ones. Pod short-pointed, light-brown. Seeds spindle-shaped.

8. **J. nodo'sus**, L. Stem erect, 6-15 inches high, slender, from a creeping slender and *tuber-bearing* rootstock, usually with 2 or 3 slender leaves. *Heads few*, 8-20-flowered, and overtopped by the involucral leaf. Flowers brown. Pod slender, taper-pointed, 1-celled. Seeds obovate, mucronate.

+ + *Stamens 3. Seeds tailed.*

9. **J. Canadensis**, J. Gay, var. **coarcta'tus**, Engel. Stems slender, 9-18 inches high, tufted, bearing 2 or 3 leaves. Panicle somewhat erect, contracted; the heads 3-5-flowered, deep-brown. Pod prismatic, abruptly pointed, deep-brown. Seeds slender, with short tails.—A very late-flowering species.

Order CIII. **PONTEDERIA'CEÆ.** (Pickerel-weed F.)

The most common representatives of this Order with us are

1. **PONTEDE'RIA**, L. Pickerel-weed.

P. corda'ta, L. A stout plant growing in shallow water, sending up a scape bearing a single large arrow-heart-shaped blunt leaf, and *a spike of violet-blue flowers with a spathe-like bract*. Perianth 2-lipped, the 3 upper divisions united, the 3 lower spreading, the whole revolute-coiled after flowering, the fleshy base enclosing the fruit. Stamens 6, 3 of them exserted on long filaments, the rest short.

2. **SCHOL'LERA**, Schreber. Water Star-grass.

S. gramin'ea, Willd. (*Heteranthera graminea*, Vahl., in Macoun's Catalogue.) A grass-like herb, wholly under water, only the small yellowish flowers reaching the surface, the latter single, from spathes. Perianth salver-shaped, regular. Stamens 3, anthers sagittate.

Order CIV. **ERIOCAULONA'CEÆ.** (Pipewort F.)

Represented with us by the genus
ERIOCAU'LON, L. Pipewort.

E. septangula're, Withering. A slender plant with a naked scape 2-6 inches high, growing in shallow water in the margins of our northern ponds. Leaves short, awl-shaped, in a tuft at the base. Flowers in a small woolly head at the summit of the scape, monœcious. Perianth double; the outer set or calyx of 2 3 keeled sepals; the corolla tubular in the sterile flowers and of 2-3 separate petals in the fertile ones. Scape 7-angled. The head (except the beard) lead-coloured.

III. GLUMA'CEOUS DIVISION.

Flowers without a proper perianth, but subtended by thin scales called *glumes*.

This Division includes two very large Orders—Cyperaceæ and Gramineæ—both of which present many difficulties to the beginner. Accordingly no attempt will be made here to enumerate and describe all the commonly occurring species of these Orders. In chapter XIV., Part I., the student will find descriptions and illustrations of several typical Grasses. We shall here, therefore, only describe two or three of the commonest representatives of the Order Cyperaceæ, so as to put the beginner in a position to continue his studies with the aid of Gray's Manual or other advanced work.

Order CV. **CYPERA'CEÆ.** (Sedge Family.)

Grass-like or rush-like herbs, easily distinguished from Grasses by the sheaths of the leaves, which in the Sedges are *closed* round the culm, not split. Flowers in spikes, each flower in the axil of a glume-like bract, either altogether without a perianth or with a few bristles or scales inserted below the ovary. Ovary 1-celled, becoming an achene (2- or 3-angled). Style 2- or 3-cleft. Stamens mostly 3, occasionally 2.

We shall describe one species of each of five genera.

1. CYPE'RUS DIANDRUS.

The plant (Fig. 256) is from 4 to 10 inches in height. The culm is *triangular*, leafy towards the base, but naked above. At the summit there is an umbel the rays of which are unequal in length, and on each ray are clustered several *flat brown-coloured* spikes, the scales of which are imbricated in two distinct rows. At the base of the umbel there are 3 leaves of very unequal length, forming a sort of involucre, and the base of each ray of the umbel is sheathed. In each spike every scale except the lowest one contains a flower in its axil. The flower (Figs. 257 and 258) is entirely destitute of perianth, and consists of 2 *stamens and an ovary surmounted by a 2-cleft style*, being consequently *perfect*.—The plant is pretty easily met with in low wet places.

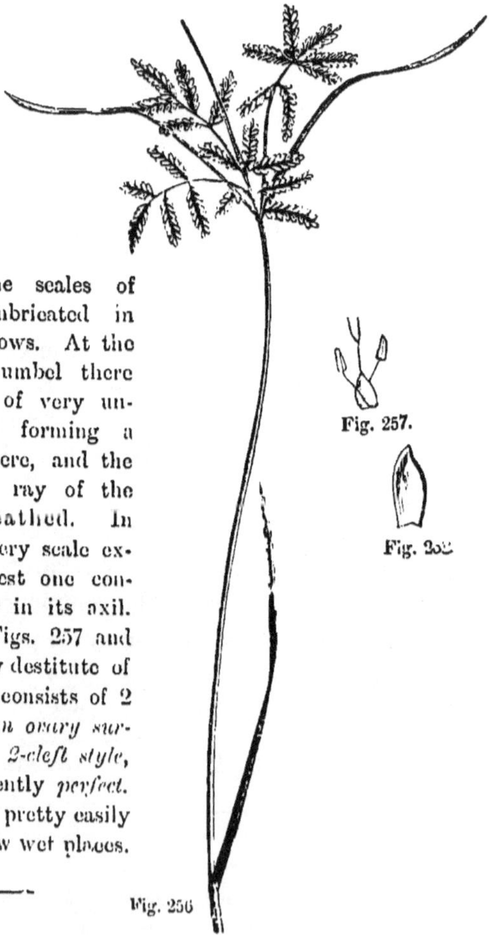

Fig. 256

Fig. 257.

Fig. 258

2. ELEOCHARIS OBTUSA.

In this plant, which grows in muddy soil in tufts 8 to 14 inches in height, there is but a single spike at the summit of each slender culm, and the scales of the spikes, instead of being imbricated in 2 rows and thus producing a flat form, are *imbricated all round.* The scales are very thin in texture, with a midrib somewhat thicker, and are usually brownish in colour. Each of them contains a perfect flower in its axil. Instead of a perianth, there are 6 or 8 *hypogynous barbed bristles.* The stamens (as is generally the case in this Order) are 3 in number, and the style is usually 3-cleft. Observe that the style is enlarged into a sort of bulb at the base, *this bulbous portion persisting as a flattish tubercle on the apex of the achene.* The culms are without leaves, being merely sheathed at the base.

3. SCIRPUS PUNGENS.

A stout marsh-plant, 2 or 3 feet high, with a sharply triangular hollow-sided culm, and bearing at the base from 1 to 3 channelled or boat-shaped leaves. The rusty-looking spikes vary in number from 1 to 6, and are in a single sessile cluster which appears to spring from the side of the culm, owing to the 1-leaved involucre resembling the culm and seeming to be a prolongation of it. Each scale of the spike is 2-cleft at the apex, and bears a point in the cleft. The flowers are perfect, with 2 to 6 bristles instead of perianth, 3 stamens, and a 2-cleft style, *but there is no tubercle on the apex of the achene.* The culms of this plant spring from stout running rootstocks.

4. ERIOPHORUM POLYSTACHYON.

A common bog-plant in the northern parts of Canada, resembling Scirpus in the details as to spikes, scales, &c., but differing chiefly in this, that the bristles of the flowers are very delicate and become very long after flowering, so that *the spike in fruit looks like a tuft of cotton.* The culm of our plant is triangular, though not manifestly so, and its leaves are hardly, if at all, channelled. The spikes are several in number, and are on nodding peduncles, and the involucre consists of 2 or 3 leaves. Culm 15 or 20 inches high.

5. CAREX INTUMES'CENS.

The species of the genus Carex are exceedingly numerous and difficult of study. The one we have selected (Fig. 259) is one of the commonest and at the same time one of the easiest to examine. In this genus the flowers are monœcious, the separate kinds being either borne in different parts of the same spike or in different spikes. The genus is distinguished from all the others of this Order by the fact of the achene *being enclosed in a bottle-shaped more or less inflated sac*, which is made by the union of the edges of two inner bractlets or scales. To this peculiar sac (Figs. 260 and 261) which encloses the achene the name *perigynium* is given. The culms are always triangular and the leaves grass-like, usually roughened on the margins and on the keel.

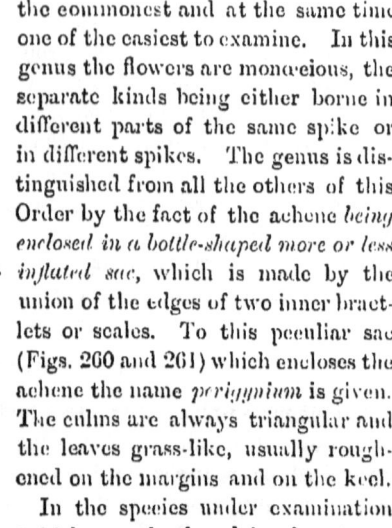

Fig. 260.

Fig. 261.

Fig. 259.

In the species under examination (which may be found in almost any wet meadow) the culm is some 18 inches high. The staminate spike (only one) is separate from and *above* the fertile ones, which are 2 or 3 in number, few- (5 to 8) flowered, and quite near together. The perigynia are very much inflated, that is, very much larger than the achenes; they are distinctly marked with many nerves, and taper gradually into a long 2-toothed beak from which protrude the 3 stigmas. The bracts which subtend the spikes are leaf-like, and extend much beyond the top of the culm.

ORDER CVI. **GRAMIN'EÆ.** (GRASS FAMILY.)

Herbs somewhat resembling those of the last Order, but the culms are hollow except at the joints, and the sheaths of the leaves are split on the opposite side of the culm from the blade.

SERIES II.

FLOWERLESS OR CRYPTOG'AMOUS PLANTS.

PLANTS not producing true flowers, but reproducing themselves by means of *spores* instead of seeds, the spores consisting merely of simple cells, and not containing an embryo.

This series is subdivided into three classes :

1. **Pteridophytes**, embracing Ferns, Horsetails, and Club-Mosses.
2. **Bryophytes**, embracing the true Mosses and Liverworts.
3. **Thallophytes**, embracing Algæ and Fungi.

Types of all of these have already been described and illustrated in Part I. We shall here enumerate the common representatives of the Pteridophytes only.

FERNS.

These beautiful plants are favourites everywhere, and we shall therefore enter into a description of their characteristics with sufficient minuteness to enable the young student to determine with tolerable certainty the names of such representatives of the Family as he is likely to meet with commonly.

In Chapter XXI. of Part I. will be found a full account of the common Polypody, with which it is assumed the student is already familiar.

Fig. 262 shows a portion of the frond of the Common Brake (Pteris aquilina). Here the frond is several times compound. The first or largest divisions to the right and left are called *pinnæ*.

170 COMMON CANADIAN WILD PLANTS.

The secondary divisions (or first divisions of the pinnæ) are the *pinnules*. The stem, as in the Polypody, and in fact in all our Ferns which have a stem at all, is a rootstock or rhizome. But here we miss the fruit-dots or sori, so conspicuous in our first example. In this case it will be found that there is a *continuous line of sporangia around the margin* of every one of the pinnules of the frond, and that the edge of the pinnule is reflexed *so as to cover the line of spore-cases*. Fig. 263 is a very much magnified view of one of the lobes of a pinnule,

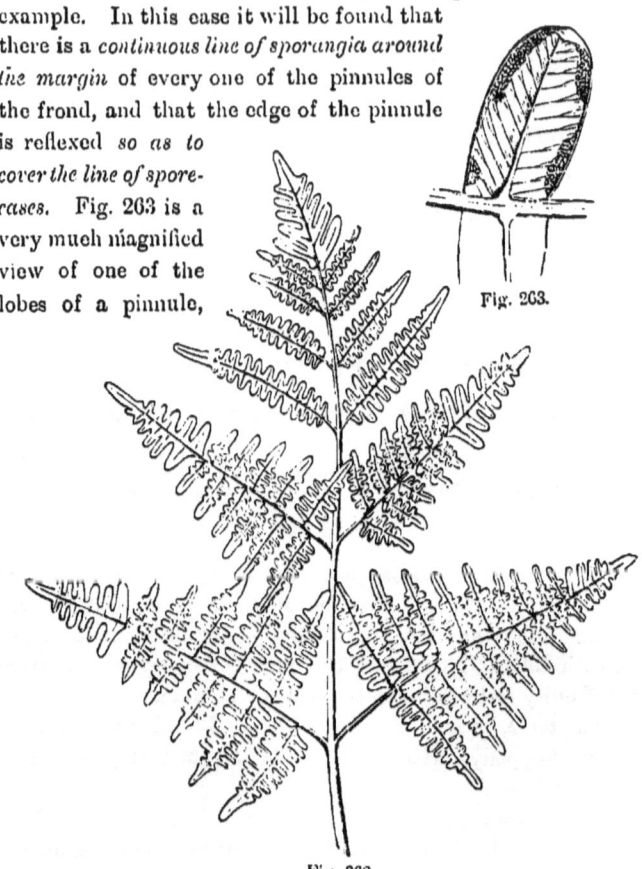

Fig. 263.

Fig. 262.

with the edge rolled back to show the sporangia. Some of the sporangia are removed to show a line which runs across the ends of the forking veins. To this the sporangia are attached. The veins, it will be seen, do not form a net-work, and so are free, as

FERNS. 171

in Polypody. Observe, then, that in Polypody the sori are not covered, whilst in Pteris the opposite is the case. The covering of the fruit-dots is technically known as the *indusium*. The individual spore-cases are alike in both plants.

Fig. 264 shows a frond of one of our commonest Shield-Ferns (Aspidium acrostichoides). It is simply pinnate. The stipe is thickly beset with rusty-looking, chaff-like scales. The veins are free, as before. The *sori* or *fruit-dots* are on the back of the upper pinnæ, but they are neither collected in naked clusters, as in Polypody, nor are they covered by the edge of the frond as in the Brake. Here each cluster has an *indusium* of its own. The indusium is round, and attached to the frond by its depressed centre (peltate). Fig. 265 shows an enlarged portion of a pinna, with the sporangia escaping from beneath the indusium. From one forking vein the sporangia are stripped off to show where they have been attached. The separate sporangia discharge their spores in the manner represented in the account of Polypody.

In some Ferns the fruit-dots are elongated instead of being round, and the indusium is attached to the frond by one *edge* only, being free on the other. Sometimes two long fruit-dots will be found side by side, the free edges of the indusia being towards each other, so that there is the *appearance* of one long fruit-dot with an indusium split down the centre.

Fig. 265.

Fig. 264.
or Sensitive Fern.

Fig. 266 represents a frond of a very common swamp Fern, Onoclea Sensibilis. It is deeply pinnatifid, and on one of the lobes the veining is represented. Here the veins are *not* free, but as they form a net-work they are said to be *reticulated*. You will look in vain on this frond for fruit-dots, but beside it grows

Fig. 266. Fig. 267. Fig. 268. Fig. 269.

another, very different in appearance,—so different that you will hardly believe it to be a frond at all. It is shown in Fig. 267. It is twice pinnate, the pinnules being little globular bodies, one of which, much magnified, is shown in Fig. 268. You may open out one of these little globes, and then you will have something like what is shown in an enlarged form in Fig. 269. It now looks more like a pinnule than when it was rolled up, and it now also displays the fruit-dots on the veins inside. Here, then, we have evidently two kinds of frond. That bearing the fruit-dots we shall call the *fertile* frond, and the other we shall call the *sterile* one. You must not look upon the pinnule in which the sori are wrapped up as an indusium. Sori which are wrapped up in this way have an indusium of their own besides, but in this plant it is so obscure as to be very difficult to observe.

The spore-cases burst open by means of an elastic ring as before.

Fig. 270 represents one of the Moonworts (Botrychium Virginicum), very common in our rich woods everywhere. Here we have a single frond, but made up manifestly of two distinct portions, the lower sterile and the upper fertile. Both portions are thrice-pinnate. The ultimate divisions of the fertile segment are little globular bodies, but you cannot unroll them as in the case of the Onoclea. Fig. 271 shows a couple of them greatly enlarged. There is a slit across the middle of each, and one of the slits is

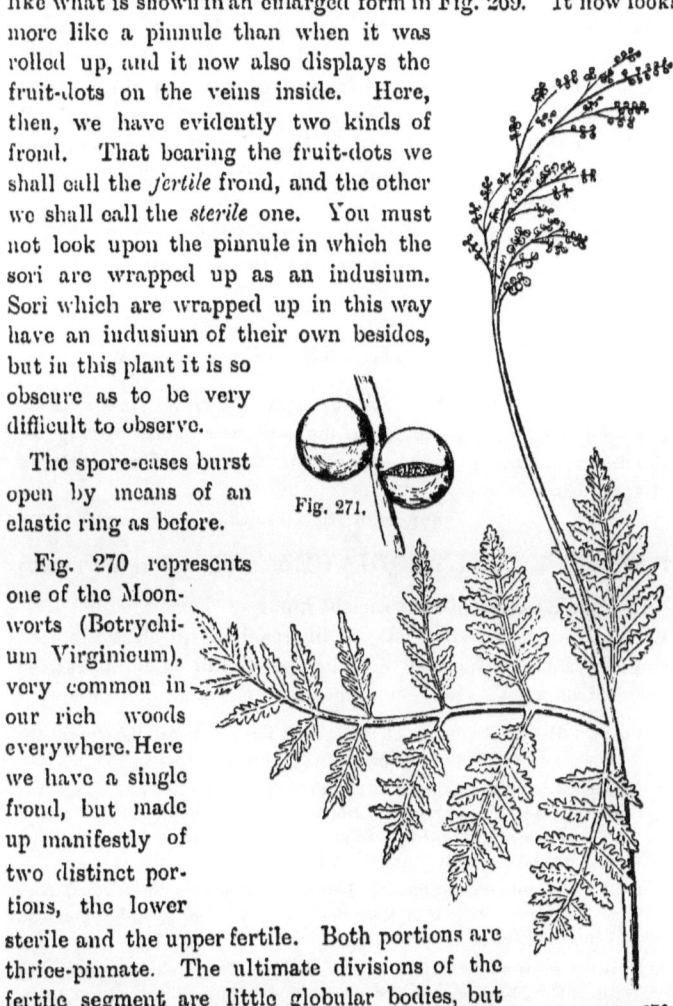

Fig. 271.

Fig. 270.

partially open, disclosing the *spores* inside. Each little globe is, in fact, a *spore-case* or *sporangium*, so that here we have something quite different from what we have so far met with. Up to this point we have found the sporaniga collected into dots or lines or clusters of some sort. In the Moonwort the sporangia are separate and naked, and instead of bursting through the action of an elastic ring, they open by a horizontal slit and discharge their spores. In other Ferns, as the Osmunda, the sporangia are somewhat similar, but burst open by a *vertical* instead of a horizontal slit.

Observe that the frond of Botrychium is *not circinate* in the bud.

We shall now proceed to describe the commonly occurring representatives of the Fern Family.

Order CVII. FIL'ICES. (Fern Family.)

Flowerless plants with distinct leaves known as *fronds*, these circinate in the bud, except in one suborder, and bearing on the under surface or margin the clustered or separate sporangia or spore-cases.

Synopsis of the Genera.

Suborder I. **POLYPODIA'CEÆ.** (The True Ferns.)

Sporangia collected into various kinds of clusters called *sori*. Each sporangium pedicelled and encircled by an elastic jointed ring, by the breaking of which the sporangium is burst and the spores discharged. Sori sometimes covered by an *indusium*.

1. **Polypo'dium.** Fruit-dots on the back of the frond near the ends of the veins. *No indusium.* Veins free. (See Fig. 231, Part I.)

2. **Adian'tum.** Fruit-dots *marginal*, the edge of the frond being reflexed so as to form an indusium. *Midrib of the pinnules close to the lower edge or altogether wanting.* Stipe *black and shining.* All the pinnules distinct and generally minutely stalked. Veins free.

3. **Pte'ris.** Fruit-dots *marginal.* Indusium formed by the reflexed edge of the frond. Midrib of the pinnules in the centre and prominent. Veins free. Stipe light-coloured. (See Fig. 262.)

4. **Pellæ'a.** Fruit-dots *marginal*, covered by a broad indusium, formed by the reflexed margin of the frond. Small ferns with once- or twice-pinnate fronds, the fertile ones very much like the sterile, but with narrower divisions. Stipe shining, purple or brown.

5. **Asple'nium.** Fruit-dots elongated, on veins on the back of the pinnules, *oblique to the midrib, but only on the upper side of the vein.* Indusium attached to the vein by one edge, the other edge free. Veins free.
6. **Woodwardia.** Fruit-dots elongated, on cross-veins *parallel to the midrib*, forming a chain-like row on each side of the latter. Indusium as in the last. *Veins reticulated.*
7. **Scolopen'drium.** Fruit-dots elongated, *occurring in pairs on contiguous veinlets*, the free edges of the two indusia facing each other, so that the sori appear to be single, with an indusium split down the centre. Veins free. *Frond simple, ribbon-shaped*, about an inch broad, generally wavy-margined.
8. **Camptoso'rus.** Fruit-dots elongated, those near the base of the midrib double, as in Scolopendrium; others single, as in Asplenium. *Fronds simple*, ½ or ¾ of an inch wide at the heart-shaped base, and tapering into a long and narrow point; growing in tufts on limestone rocks, and commonly rooting at the tip of the frond, like a runner. Veins reticulated.
9. **Phegop'teris.** Fruit-dots roundish, on the back (not at the apex) of the veinlet, rather small. *Indusium obsolete or none.* Veins free. *Fronds triangular in outline*, in one species twice-pinnatifid, with a winged rhachis, and in the other in three petioled spreading divisions, the divisions once- or twice-pinnate.
10. **Aspid'ium.** Fruit-dots round. Indusium evident, flat, orbicular or kidney-shaped, fixed by the centre, opening all round the margin. Veins free. Generally rather large Ferns, from once- to thrice-pinnate. (See Fig. 264.)
11. **Cystop'teris.** Fruit-dots round. Indusium not depressed in the centre, but rather raised, attached to the frond not by the centre, but by the edge partly under the fruit-dot, and generally breaking away on the side towards the apex of the pinnule, and becoming reflexed as the sporangia ripen. Fronds slender and delicate, twice- or thrice-pinnate.
12. **Struthiop'teris.** Fertile frond much contracted and altogether unlike the sterile ones, the latter very large and growing in a cluster with the shorter fertile one in the centre. Rootstock very thick and scaly. Fertile fronds simply pinnate, the margins of the pinnæ rolled backwards so as to form a hollow tube containing the crowded sporangia. Very common in low grounds.
13. **Onocle'a.** Fertile and sterile fronds unlike. (See Figs. 266, 267, 268, 269, and accompanying description.)
14. **Dickso'nia.** Fruit-dots round, *very small, each on a recurved toothlet* on the upper margin of the lobes of the pinnules, usually one to each lobe. Sporangia on an elevated globular receptacle, and enclosed in a cup-shaped indusium open at the top and partly adherent to the reflexed toothlet of the frond. Fronds minutely glandular and hairy, 2-3 feet high, ovate-lanceolate in outline, pale green, very thin, without chaff.

SUBORDER II. **OSMUNDA'CEÆ.**

Sporangia naked, globular, pedicelled, *reticulated*, opening by a vertical slit.

15. **Osmun'da.** Fertile fronds or *fertile portions* of the frond much contracted, bearing naked sporangia, which are globular, short-pedicelled, and opening by a *vertical* slit to discharge the spores. Frond tall and upright, once- or twice-pinnate, from thick rootstocks.

SUBORDER III. **OPHIOGLOSSA'CEÆ.**

Sporangia naked, *not reticulated*, opening by a horizontal slit. *Fronds not circinate in the bud.*

16. **Botrych'ium.** Sporangia in compound spikes, *distinct*, opening by a horizontal slit. Sterile part of the frond compound. Veins free. (See Figs. 270 and 271.)

17. **Ophioglos'sum.** Sporangia *coherent* in 2 ranks on the edges of a simple spike. Sterile part of the frond simple. Veins reticulated.

1. POLYPO'DIUM, L. POLYPODY.

P. vulga're, L. Fronds evergreen, 4-10 inches long, deeply pinnatifid, the lobes obtuse and obscurely toothed. Sori large.—Common on shady rocks.

2. ADIAN'TUM, L. MAIDENHAIR.

A. peda'tum, L. Stipe upright, black and shining. The frond forked at the top of the stipe, the two branches of the fork recurved, and each bearing on its inner side several slender spreading divisions, the latter with numerous thin pinnatifid pinnules which look like the *halves* of pinnules, owing to the midrib being close to the lower edge. Upper margin of the pinnules cleft.—Common in rich woods.

3. PTE'RIS, L. BRAKE. BRACKEN.

P. aquili'na, L. Stipe stout and erect. Frond large and divided into 3 large spreading divisions at the summit of the stipe, the branches twice-pinnate, the pinnules margined all round with the indusium.—Common in thickets and on dry hillsides.

4. PELLÆ'A, Link. CLIFF-BRAKE.

1. **P. gra'cilis,** Hook. Fronds 3-6 inches high, slender, of few pinnae, the lower ones once- or twice-pinnatifid into 3-5 divisions,

those of the fertile fronds narrower than those of the sterile ones. Stipe polished, *brownish*, darker at the base.—Shady limestone rocks; not common.

2. **P. atropurpu′rea**, Link. Larger than the last, 6–15 inches high, the stipe *dark-purple* and shining. Frond pale, once- or (below) twice-pinnate, the divisions broadly linear or oblong (or the sterile sometimes oval), stalked at the base. Stipes clustered.—Dry rocks.

5. ASPLE′NIUM, L. SPLEENWORT.

1. **A. Trichom′anes**, L. A very delicate little fern growing in tufts on shaded cliffs. Fronds 3–6 inches long, linear in outline, pinnate, the little pinnæ oval and unequal-sided, about $\frac{1}{5}$ of an inch long. The stipes thread-like, purplish-brown and shining. This species is evergreen.

2. **A. thelypteroi′des**, Michx. Fronds 2–3 feet high, *pinnate*, the pinnæ linear-lanceolate in outline, 3–5 inches long, *deeply pinnatifid*, each of the crowded lobes bearing 3–6 pairs of oblong fruit-dots.—Rich woods.

3. **A. angustifo′lium**, Michx. Fronds simply pinnate, somewhat resembling Aspidium acrostichoides, *but very smooth and thin*, and larger. Pinnæ crenulate, short-stalked. Fruit-dots linear, crowded.—Rich woods; not common.

4. **A. Filix-fœ′mina**, Bernh. Fronds 1–3 feet high, broadly lanceolate in outline, *twice-pinnate*, the pinnæ lanceolate in outline, and the pinnules confluent by a narrow margin on the rhachis of the pinna, doubly serrate. *Indusium curved*, often shaped something like a horse-shoe, *owing to its crossing the vein and becoming attached to both sides of it.*—Rich woods.

6. WOODWARD′IA, Smith. CHAIN FERN.

W. Virgin′ica, Smith. Fronds 2–3 feet high, pinnate; pinnæ lanceolate, pinnatifid. Veins forming a single row of meshes next the midrib.—Wet swamps.

7. SCOLOPEN′DRIUM, Smith. HART'S TONGUE.

S. vulga′re, Smith. Frond simple, bright green, a foot or more in length, and an inch or more in width.—Shaded ravines and limestone cliffs; not very common.

8. CAMPTOSO'RUS, Link. WALKING-LEAF.

C. rhizophyl'lus, Link. A curious little fern, growing in tufts on shaded limestone rocks. Frond simple, with a very long narrow point.—Not very common.

9. PHEGOP'TERIS, Fée. BEECH FERN.

1. **P. polypodioi'des**, Fée. Fronds triangular, longer than broad, 4-6 inches long, hairy on the veins, twice-pinnatifid, *the rhachis winged*. The pinnæ sessile, linear-lanceolate in outline, *the lowest pair deflexed and standing forwards*. Fruit-dots small and *all* near the margin. Stipes rather longer than the fronds, from a slender, creeping rootstock.—Apparently not common, but growing in rich woods near Barrie, Ont.

2. **P. hexagonop'tera**, Fée. Fronds triangular, generally broader than long, 7-12 inches broad. Pinnæ lanceolate; the lowest very large, their divisions elongated and pinnatifid, the basal divisions decurrent on the main rhachis and forming a many-angled wing. Fruit-dots not exclusively near the margin.—Rich woods.

3. **P. Dryop'teris**, Fée. Fronds broadly triangular in outline, primarily divided into 3 triangular spreading petioled divisions, smooth, the three divisions once- or twice-pinnate. Fronds from 4 to 6 inches wide. Fruit-dots near the margin.—Rich woods; common. Whole plant delicate, and light green in colour.

10. ASPID'IUM, Swartz. SHIELD FERN. WOOD FERN.

* Stipes not chaffy.

1. **A. thelyp'teris**, Swartz. Fronds tall and narrow, lanceolate in outline, pinnate, the pinnæ deeply pinnatifid, nearly at right angles to the rhachis, linear-lanceolate in outline, *the margins of the lobes strongly revolute in fruit*. Stipe over a foot long, and usually longer than the frond.—Common in low, wet places.

2. **A. Noveboracen'se**, Swartz. *Fronds much lighter in colour than the preceding*, tapering towards both ends, pinnate, the pinnæ deeply pinnatifid, much closer together than in No. 1, and not at right angles with the rhachis. Veins simple. Lower pinnæ short and deflexed.—Swamps.

* * *Stipes chaffy.*

3. **A. spinulo'sum**, Swartz. Stipes *slightly* chaffy or scaly. Fronds large, ovate-lanceolate in outline, *twice-pinnate*, the pinnules deeply pinnatifid (*nearly pinnate*), and spiny-toothed. Pinnæ triangular-lanceolate in outline. The variety **intermedium**, which is very common in Canadian woods, has the few scales of the stipe pale brown *with a dark centre*, and *the lower pinnæ unequal-sided*. Var. **Boottii** has the scales of the stipe pale brown, the frond elongated-oblong or elongated-lanceolate and pinnules less dissected.

4. **A. crista'tum**, Swartz. Stipes chaffy with broad scales. Fronds large, linear-lanceolate in outline, once-pinnate, the pinnæ deeply pinnatifid, the upper ones triangular-lanceolate in outline, the lower considerably broader, the lobes cut-toothed. Fruit-dots large and conspicuous, *half way between the midrib of the lobe and the margin.*—Swamps.

5. **A. Goldia'num**, Hook. A fine fern, the large fronds growing in a circular cluster from a chaffy rootstock. Frond ovate or ovate-oblong in outline, once-pinnate, the pinnæ deeply pinnatifid, 6-9 inches long, *broadest in the middle*, the lobes slightly scythe-shaped, finely serrate. Fruit-dots large, *near the midrib of the lobe.*—Rich moist woods.

6. **A. margina'le**, Swartz. Stipes very chaffy at the base. Fronds ovate-oblong in outline, twice-pinnate, the pinnæ lanceolate in outline, broadest above the base. Pinnules crenate-margined. *Fruit-dots large, close to the margin.*—Rich woods, mostly on hill-sides.

7. **A. acrostichoi'des**, Swartz. (See Figs. 264 and 265, and accompanying description.)—Rich woods, everywhere.

8. **A. Lonchi'tis**, Swartz. Not unlike No. 7, but the fronds are *narrower and longer*, more rigid and with hardly any stipe. Pinnæ densely spinulose-toothed.—Apparently not common, but plentiful in rocky woods west of Collingwood, Ont.

11. CYSTOP'TERIS, Bernhardi. BLADDER FERN.

1. **C. bulbif'era**, Bernh. Frond large (1-2 feet), narrow and very delicate, twice-pinnate, the pinnæ nearly at right angles to

the rhachis. Rhachis and pinnæ *usually with bulblets beneath.* Pinnules toothed.—Shady, moist ravines.

2. **C. fra'gilis**, Bernh. Frond only 4–8 inches long, with a stipe of the same length, twice- or thrice-pinnate. *Rhachis winged.*—Shady cliffs.

12. STRUTHIOP'TERIS, Willd. OSTRICH FERN.

S. German'ica, Willd. (*Onoclea Struthiopteris,* Hoff.) Sterile fronds with the lower pinnæ gradually much shorter than the upper ones. Pinnæ deeply pinnatifid.—Common in low, wet grounds along streams.

13. ONOCLE'A, L. SENSITIVE FERN.

O. sensib'ilis, L. (See Figs. 266, 267, 268 and 269, and accompanying description.)—Common in wet grounds along streams.

14. DICKSO'NIA, L'Her. DICKSONIA.

D. punctilo'bula, Kunze. Pleasantly odorous.—Moist, shady places.

15. OSMUN'DA, L. FLOWERING FERN.

1. **O. rega'lis,** L. (FLOWERING FERN.) Fronds twice-pinnate, *fertile at the top,* very smooth, pale green. Sterile pinnules oblong-oval, finely serrate towards the apex, 1–2 inches long, either sessile or short-stalked, usually oblique and truncate at the base.—Swamps, along streams and lake-margins.

2. **O. Claytonia'na,** L. Fronds large, once-pinnate, pale green, densely white-woolly when unfolding from the bud, *with fertile pinnæ among the sterile ones.* Pinnæ deeply pinnatifid, the lobes entire.—Low grounds.

3. **O. cinnamo'mea,** L. (CINNAMON FERN.) *Fertile fronds distinct from the sterile ones,* contracted, twice-pinnate, covered with cinnamon-coloured sporangia. Sterile fronds rusty-woolly when young, smooth afterwards, once-pinnate, the pinnæ deeply pinnatifid. The long, sterile fronds in a cluster, with the fertile ones in the centre.—Low grounds.

16. BOTRYCH'IUM, Swartz. MOONWORT.

1. **B. Virgin'icum,** Swartz. (See Figs. 270 and 271, and accompanying description.) Rich woods everywhere.

2. **B. lunarioi'des**, Swartz, is occasionally found. It is easily distinguished from No. 1 by the sterile portion of the frond being long-petioled instead of sessile.

17. OPHIOGLOS'SUM, L. ADDER'S TONGUE.

O. vulga'tum, L. Sterile part of the frond ovate or elliptical-oblong, 2-3 inches long, rather fleshy, sessile, near the middle of the stalk; the latter 6-12 inches high.—Bogs and grassy meadows.

ORDER CVIII. EQUISETA'CEÆ.
(HORSETAIL FAMILY.)

The only genus of the Order is

EQUISE'TUM, L. HORSETAIL. SCOURING RUSH

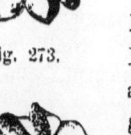

Fig. 273.

Fig. 274.

Fig. 272 is a view of the fertile stem of **Equise'tum arvense**, the COMMON HORSETAIL, of about the natural size. It may be observed early in spring almost anywhere in moist sandy or gravelly soil. It is of a pale brown colour, and in place of leaves there is at each joint a sheath split into several teeth. At the summit of the stem is a sort of conical catkin, made up of a large number of six-sided bodies, each attached to the stem by a short pedicel. Each of these six-sided bodies turns out on examination to be made up of six or seven sporangia or spore-cases, which open down their inner margins to discharge their spores. Figs. 273 and 274 are enlarged outer and inner views of one of them. The spores themselves are of a similar nature to those of the Ferns, and reproduction is carried on in the same manner; but each spore of the Horsetail is furnished with four minute tentacles which closely envelope it when moist, and uncoil themselves when dry. The use of these tentacles is doubtless to assist in the escape and dispersion of the spores.

Fig. 272.

The fertile stems will have almost withered away by the time the sterile ones appear. These latter are of the same thickness as the fertile ones, but they are very much taller and are green in colour. Observe, also, the grooving of the sterile stem, and the whorls of 4-angled branches produced at the nodes.

E. limo'sum, L., produces its spores in the summer. The stems are all of one kind, or at all events are produced contemporaneously, and after fruiting produce upright branches. They are, also, *slightly* many-furrowed. The sheaths have commonly about 18 dark-brown stiff short teeth.—In shallow water.

The two species just described are annual; the following are evergreen, surviving the winter. The terminal spike is tipped with a small rigid point.

E. hyema'le, L. (SCOURING RUSH.) Stems stout and tall. Sheaths elongated, with a black girdle above the base, and about 20 narrow linear teeth, 1-keeled at the base, and with awl-shaped deciduous points.—Wet banks.

E. variega'tum, Schleicher. Stem *slender*, in tufts, with 5-10 grooves, ascending, 6-18 inches high. Sheaths green, variegated with black above, 5-10-toothed.—Shores and river-banks.

E. scirpoi'des, Michx. Stems *slender, very numerous in a tuft*, filiform, 3-6 inches high, curving, mostly 6-grooved. *Sheaths 3-toothed.*—Wooded hill-sides.

ORDER CIX. **LYCOPODIA'CEÆ.** (CLUB-MOSS F.)

Chiefly moss-like plants; often with long running and branching stems, the sporangia solitary in the axils of the mostly awl-shaped leaves.

Synopsis of the Genera.

1. **Lycopo'dium.** *Spore-cases of one kind only*, in the axils of the upper awl-shaped leaves; 2-valved, kidney-shaped. Chiefly evergreen plants. (See Part I., sections 332-335.)

2. **Selaginel'la.** *Spore-cases of two kinds:* one like those of Lycopodium, containing very minute spores, the other 3-4-valved, and containing a few large spores. The two sorts intermingled, or the latter in the lower axils of the spike. Little moss-like tufted evergreens.

LYCOPODIACEÆ.

1. LYCOPO'DIUM, L., Spring. CLUB-MOSS.

1. **L. luci'dulum**, Michx. Stems 4-8 inches long, tufted, 2 or 3 times forking. The leaves forming the spike not different from the others on the stem; all spreading or reflexed, sharp-pointed, serrulate, dark green and shining.—Cold, moist woods.

2. **L. anno'tinum**, L. Stems creeping, 1-4 feet long. Branches 4-9 inches high, once- or twice-forked. Spike sessile, the leaves of it yellowish and scale-like, ovate or heart-shaped, the others spreading or reflexed, rigid, pointed, nearly entire, pale green.—Cold woods.

3. **L. dendroi'deum**, Michx. (GROUND PINE.) Rootstock creeping underground, nearly leafless. Stems much resembling little hemlocks, 6-9 inches high; numerous fan-like spreading branches with shining lanceolate entire leaves. Spikes nearly as in No. 2, 4-10 on each plant.—Moist woods.

4. **L. clava'tum**, L. (CLUB-MOSS.) Stem creeping or running extensively. *Spikes mostly in pairs*, raised on a slender peduncle (4-6 inches long). Leaves linear, awl-shaped, *bristle-tipped.*—Dry woods.

5. **L. complana'tum**, L. Stem creeping extensively. *Branches flattened*, forking above, the branchlets crowded. Leaves awl-shaped, small, in 4 ranks.—Dry woods; mostly with evergreens.

2. SELAGINEL'LA, Beauv., Spring.

S. rupes'tris, Spring. A little moss-like evergreen, growing on exposed rocks in dense tufts 1-3 inches high. Leaves awl-shaped, with a grooved keel, and tipped with a bristle. Spikes 4-cornered.

AN APPENDIX

CONTAINING

Descriptions of Plants found in the Maritime Provinces of Canada

NOT HEREIN BEFORE DESCRIBED.

The student will be able to follow the plant to its proper order, in nearly every case, and generally to its genus, by the use of the preceding Flora. If he cannot find the *specific* description there, he will then turn to the Appendix.

APPENDIX.

ORDER I. **RANUNCULA'CEÆ.** (CROWFOOT FAMILY.)

CLEM'ATIS.

C. verticilla'ris, DC. Shrubby climber. Peduncles bearing single large flowers, with thin, wide-spreading, bluish-purple sepals. Tails of the achenes plumose. Leaves of three leaflets, which are entire, or sparingly toothed or lobed.—Rocky places.

ANEMONE.

A. parviflo'ra, Michx. (SMALL-FLOWERED ANEM'ONE.) Stem 3-12 inches high, one-flowered. Sepals 5 or 6, white. Involucre 2-3-leaved far below the flower. Head of carpels woolly, globular. Root-leaves small, 3-parted, their divisions crenately lobed.—Rocky river-margins.

A. multif'ida, DC. (MANY-CLEFT A.) Silky-hairy. Principal involucre 2-3-leaved, bearing one naked and one or two 2-leaved peduncles. Leaves of the involucre short-petioled, twice or thrice 3-parted and cleft, their divisions linear. Sepals red, greenish-yellow, or whitish. Head of carpels spherical or oval, woolly.— Rocky river-margins, etc.

RANUNCULUS.

R. Cymbala'ria, Pursh. (SEA-SIDE CROWFOOT.) Low, smooth, spreading by runners which take root at the joints. Leaves long-petioled, roundish, crenate, rather fleshy. Petals 5-8, yellow. Carpels striate, in an oblong head.—Sea-shore and beside brackish streams and springs.

AQUILEGIA.

A. vulga'ris, L. The Garden Columbine is found escaped from cultivation in some places. It has *hooked spurs*, whereas those of A. Canadensis are nearly straight.

ORDER V. **BERBERIDA'CEÆ.** (BARBERRY FAMILY.)

BERBERIS.

B. vulga'ris, L. (COMMON BARBERRY.) Shrub. Leaves on the fresh shoots of the season mostly reduced to branched spines, from whose axils proceed the next year close clusters of obovate oblong, bristly-toothed leaves, with short, *jointed* petioles, and many-flowered racemes. Sepals, petals and stamens 6 each. Outside of sepals are 2-6 bractlets. Petals yellow. Fruit an oblong, sour, scarlet berry.—Cultivated grounds.

ORDER X. **CRUCIF'ERÆ.** (CRESS FAMILY.)

ARABIS.

A. petræ'a, Lam. Petals rose-color or whitish. Pods shorter and less flat than A. lyrata. Seeds in one row in each cell. Leaves spathulate or oblong, entire or sparingly toothed.—Rocks.

DRABA.

D. inca'na, L. Hoary-pubescent. Leaves lanceolate or oblanceolate to ovate, entire or sparingly toothed. Pods oblong to lanceolate usually straight, on short erect pedicels. Style short or none.—Dry rocks.

RAPHANIS.

R. Raphanis'trum, L. (WILD RADISH.) Pods linear or oblong, tapering, indehiscent, long-beaked, necklace form when ripe. Style long. Cotyledons conduplicate. Leaves rough, lyrate. Petals yellow, veiny, turning whitish or purplish.—An introduced weed.

ORDER XII. **VIOLA'CEÆ.** (VIOLET FAMILY.)

VIOLA.

V. lanceola'ta, L. (LANCE-LEAVED VIOLET.) Stemless. Flowers white. Petals beardless. Leaves lanceolate, erect, tapering into a long, margined petiole, nearly entire.—Damp ground.

V. primulæfo'lia, L. (PRIMROSE-LEAVED V.) Stemless. Flowers white, lateral petals usually somewhat bearded. Leaves ovate or oblong, heart-shaped, or abrupt at the base.—Damp or dry ground.

Order XVI. CARYOPHYLLA'CEÆ. (Pink Family.)

STELLARIA.

S. lon'gipes, Goldie. (Long-stalked Stitchwort.) Leaves somewhat rigid, ascending, lanceolate, acute, broadest at the base. Cyme terminal, few-flowered, the long pedicels erect, scaly-bracted. Petals longer than the sepals. Seeds smooth.

S. gramin'ea, L. Like the last, but the leaves *broadest above the base, the pedicels widely spreading*, and the seeds strongly but finely rugose. (Int. from Eu.)

S. uligino'sa, Murr. (Swamp S.) Stems weak, decumbent or diffuse. *Leaves lanceolate, or oblong*, veiny. Petals and ripe pods *as long as the sepals*. Seeds roughened. *Cymes naked*, becoming lateral.—Swamps and rills.

S. crassifo'lia, Ehrh. Stems diffuse or erect, weak. *Leaves rather fleshy*, lanceolate to oblong, those of flowering branches smaller and thinner. Petals longer than sepals or wanting. *Seeds rugose-roughened*. Flowers terminal or in the forks of stem or branches.—Wet places.

S. borealis, Bigelow. (Northern S.) Stem erect or spreading, weak, forking. Leaves broadly-lanceolate to ovate-oblong. Petals 2-5, shorter than the sepals or wanting. Cyme leafy. *Seeds smooth*.—Wet places.

S. humifu'sa, Rottb. Low, spreading or creeping. Leaves *fleshy*, ovate or oblong. Pedicels axillary or terminal, on leafy stems or branches. *Petals a little longer than the sepals*. Seeds smooth.

SAGI'NA.

S. procum'bens, L. (Pearlwort.) A low, matted herb with narrowly linear leaves. Flowers small, terminal, with their parts in fours, rarely in fives. *Petals shorter than the ovate, obtuse sepals, or none.* Pod many-seeded, 4-5-valved. *Top of peduncle often bent into a hook.*—Damp places.

S. nodo'sa, Fenzl. A low, tufted herb. Lower leaves thread-form; the upper short, awl-shaped, *with clusters of minute ones in their axils. Parts of flower in fives*, the stamens sometimes ten. *Petals much longer than sepals.* Flowers terminal. Pods as in S. procumbens.—Wet sandy shores.

BUDA. SAND-SPURREY.

Sepals 5. Petals 5, entire. Stamens 2-10. Styles usually 3. Pod many-seeded, usually 3-valved. Low herbs with membranaceous stipules, and filiform or linear, opposite leaves.

B. ru'bra, Dumort. (SPERGULA'RIA RU'BRA, PRESL.) Leaves linear, flat, hardly fleshy. Stipules lanceolate. Stems usually glandular-pubescent near the summit. Calyx rather longer than the pink-red corolla, and small pod. Seeds rough with projecting points.—Dry sandy ground.

B. mari'na, Dumort. (SPERGULARIA SALINA, PRESL.) More fleshy than B. rubra, usually pubescent. Stipules ovate. Leaves terete. Sepals a little shorter than the pod. Petals pale. Seeds usually roughened with points.—Sea-coast.

B. borea'lis, Watson. (SPERGULARIA MEDIA, PRESL.) Much branched, glabrous. Petals white. Pod about twice as long as the sepals, nearly or quite smooth.—Sea-coast.

SPER'GULA.

S. arven'sis, L. (CORN SPURREY.) An annual herb resembling a Buda, with numerous thread-like leaves in whorls. Flowers white in panicled cymes. Pod 5-valved.—Grain fields. (Int. from Eu.)

ORDER XVII. **PORTULACA'CEÆ**. (PURSLANE FAMILY.)

MONTIA.

M. fonta'na, L. A small, spreading, somewhat fleshy herb, with opposite, spathulate leaves. Sepals 2. Petals 5, three of them somewhat smaller. Stamens usually 3, borne on the claws of the smaller petals. Pod 3-valved and 3-seeded.—Wet places.

ORDER XIX. **TILIA'CEÆ**. (LINDEN FAMILY.)

TILIA.

T. Europae'a, L. The European Linden is planted as a shade tree in some places. It is easily distinguished by the absence of *petal-like scales among the stamens*, which are present in the native Linden.

APPENDIX. 191

ORDER XXXI. **LEGUMINO'SÆ.** (PULSE FAMILY.)

TRIFOLIUM.

T. hybridum, L. (ALSIKE CLOVER.) Resembles T. repens, but the flowers are rose-tinted, and the stems erect or ascending. (Int. from Eu.)

OXYTROPIS.

O. campes'tris, DC. (OXYTROPIS.) Resembling Astragalus, but the keel or the corolla *tipped with a sharp point*. A low plant, almost stemless, with a hard thick root or root-stock. Flowers white or yellowish, often tinged with purple or violet, in short spikes on naked scapes. Pods ovate or oblong, of a *thin or papery* texture. Leaves pinnate, of many leaflets.—Gravelly river-margins.

HEDYSARUM.

H. borea'le, Nutt. (HEDYSARUM.) Herb with odd-pinnate leaves. Leaflets 13-21, oblong or lanceolate. Stipules scaly, united opposite the petiole. Flowers in racemes, purple. Standard shorter than the keel. Calyx 5-cleft, with nearly equal awl-shaped lobes. Pod of 3-4 flattened roundish joints, which are easily separated.—Rocky banks.

VICIA.

V. tetrasper'ma, L. Peduncles slender, 1-2-flowered. Calyx-teeth unequal. Corolla whitish. Pod 4-seeded, smooth. Leaflets about 4 pairs. (Int. from Eu.)

ORDER XXXII. **ROSA'CEÆ.** (ROSE FAMILY.)

GEUM.

G. macrophyl'lum, Willd. Bristly-hairy, stout. Root-leaves interruptedly pinnate, *with a very large round-heart-shaped terminal leaflet*. Stem-leaves with 2-4 minute lateral leaflets, the terminal 3-cleft, with *wedge-form rounded lobes*. Petals yellow, longer than the calyx. *Receptacle nearly naked.*

POTERIUM.

P. Canaden'se, Benth. and Hook. (CANADIAN BURNET.) Flowers in a long cylindrical spike, white. Calyx-tube top-shaped, persistent, with 4 broad, deciduous, petal-like, spreading lobes. Petals wanting. Stamens 4, exserted. Leaves unequally pinnate with numerous serrate leaflets, heart-shaped at base.—Bogs and wet grounds.

RU'BUS.

R. Chamæmo'rus, L. (CLOUD BERRY. BAKED-APPLE BERRY.) A low herb with *diœcious* flowers. *Stem simple* without prickles, 2-3-leaved, bearing one large white flower. Leaves simple, kidney-form, 5-lobed, serrate.—Sphagnous swamps.

ROSA.

R. hu'milis, Marsh. Low, more slender and less leafy than R. lucida, with *straight slender spines*. *Stipules narrow*. Leaflets thin and pale. *Outer sepals always more or less lobed.*—Mostly in sandy soil.

R. nit'ida, Willd. Low. Stem and branches usually thickly covered with *prickles interspersed with straight slender spines*. Stipules mostly dilated. Leaflets bright green and shining, mostly narrowly oblong. Flowers generally solitary. *Sepals entire.*—Margins of swamps.

ORDER XXXIII. **SAXIFRAGACEÆ.** (SAXIFRAGE F.)

PARNAS'SIA.

P. parviflo'ra, DC. *Petals but little longer than the sepals.* Sterile filaments *about 7 in each set*. Leaves ovate or oblong.

SAXIF'RAGA.

S. Aizo'on, Jacq. Scape 5-10 inches high. Leaves *thick, spathulate, with white, finely-toothed margins*. Petals cream-color, obovate, often spotted at the base.—Moist rocks.

ORDER XXXIV. **CRASSULACEÆ.** (ORPINE FAMILY.)

SE'DUM.

S. Tele'phium, L. (LIVE-FOR-EVER.) Stems tall and stout. Leaves oval, toothed. Flowers in compound cymes, *petals purple*. *Sepals, petals, and carpels 5 each*. Stamens 10. (Int. from Eu.)

APPENDIX. 193

S. Rhodi'ola, DC. (ROSE-ROOT.) Stems 5-10 inches high. Flowers diœcious, greenish-yellow or purplish. *Stamens mostly 8, other parts in fours.*—Rocky shores.

ORDER XXXVI. **HALORA'GEÆ.** (WATER-MILFOIL F.)

PROSERPINA'CA.

P. palustris, L. (MERMAID WEED.) Low herb. Stem creeping at base. Leaves alternate, lanceolate, sharply serrate. Petals none. Stamens 3. Fruit nut-like, 3-seeded.—Wet swamps.

MYRIOPHYL'LUM.

M. tenel'lum, Bigel. Flowering stems nearly leafless. Bracts small, entire. Flowers alternate, monœcious. Stamens 4.—Borders of ponds.

CALLIT'RICHE.

C. ver'na, L. Amphibious. Floating leaves obovate, tufted; submersed leaves linear. Flowers monœcious, axillary, usually between a pair of bracts. Sterile flower *a single stamen;* fertile flower *a single pistil* with a 4-celled ovary. *Leaves beset with stellate scales.*

C. autumna'lis. Growing under water. May be distinguished from C. verna by its leaves being *retuse and all linear from a broader base,* and its flowers *without bracts.*

ORDER XLII. **UMBELLIF'ERÆ.** (PARSLEY FAMILY.)

LIGUS'TICUM.

L. Scot'icum, L. (SCOTCH LOVAGE.) Stems smooth, from large aromatic roots. Leaves twice ternate, coarsely toothed. Flowers white. Fruit with prominent acute ribs having broad spaces between.—Salt marshes and sea-shore.

ARCHANGEL'ICA.

A. Gme'lini, DC. Stem slightly downy at the top. Involucels about as long as the umbellets. Plant but little aromatic.—Rocky coasts.

ORDER XLV. CAPRIFOLIA'CEÆ. (HONEYSUCKLE F.)

VIBUR'NUM.

V. pauciflo'rum, Pylaie. A low shrub. Leaves 5-ribbed at the base, serrate, with 3 short lobes at the summit. Cyme few-flowered. Stamens shorter than the corolla. Fruit red, sour, with a very flat stone.—Cold woods.

ORDER XLVI. RUBIA'CEÆ. (MADDER FAMILY.)

HOUSTO'NIA.

H. cæru'lea, L. (BLU'ETS. INNOCENCE.) A slender herb with erect stems. A single flower on each slender peduncle. Leaves oblong-spathulate. Corolla light blue to nearly white, with a yellowish eye and a long tube.—Moist grassy places.

ORDER XLIX. COMPOSITÆ. (COMPOSITE FAMILY.)

CENTAUREA.

C. ni'gra, L. (KNAPWEED.) Scales of the involucre *with a black hair-like fringe. Rays wanting.* Leaves lanceolate, entire, or the lower coarsely toothed, rough.—Waste places. (Int. from Eu.)

TANACETUM.

T. Huronen'se, Nutt. Hairy when young. Heads usually few, and much larger than in T. vulgare. *Pistillate flowers flattened*, instead of terete as in T. vulgare.—River margins.

ARTEMIS'IA.

A. cauda'ta, Michx. *Not hoary.* Leaves pinnately dissected into thread-form divisions. Racemes forming an elongated panicle. *Disk-flowers perfect but sterile, marginal flowers fertile.*—Sandy soil.

A. bien'nis, Willd. (BIENNIAL WORMWOOD.) Glabrous. Lower leaves twice pinnately-parted, the upper pinnatifid, lobes acute. Heads in short axillary spikes, together forming a clustered, leafy panicle. *Flowers all fertile.*—Gravelly banks. (Int. from Western States.)

BELLIS.

B. perennis. The true Daisy, a native of the Old World, is a low stemless herb. It is an uncommon garden escape. The heads are many flowered with numerous pistillate rays. The scales of the involucre equal, in about 2-rows, herbaceous. Receptacle conical. Pappus wanting.

GNAPHALIUM.

G. sylvat'icum, L. Erect, usually 9-12 inches high. *Leaves linear.* Heads axillary, nearly sessile, forming an erect leafy spike. Scales obtuse *with a brown bar across each near the top.*

SOLIDAGO.

S. puber'ula, Nutt. *Stem and panicle minutely hoary.* Stem-leaves lanceolate, acute, tapering to the base, somewhat smooth, the lowermost spathulate, sparingly toothed. Heads not large, crowded in compact short racemes, which form a long, dense, terminal panicle. Rays 10-14.—Barren soil.

S. uligino'sa, Nutt. (*S. stricta, Ait.*) Smooth. Stem simple, strict. Leaves lanceolate, pointed, the lower tapering into winged petioles, finely but sparingly serrate, or entire. Racemes crowded and appressed in a close wand-like panicle. Heads middle-sized. Rays 5-6, small.—Peat bogs and wet places.

S. macrophyl'la, Pursh. (*Solidago thyrsoidea, E. Meyer.*) Stems stout, simple, pubescent near the summit. Leaves thin, ovate, with *sharp projecting teeth*, the lower ones *abruptly contracted into long margined petioles. Heads large*, in an oblong raceme, loose and thin, long-pointed. Rays 8-10, *long.*—Wooded hillsides.

S. semper'virens, L. Stem stout. Leaves *long*, lanceolate, *thickish*, smooth, entire, obscurely 3-nerved. Racemes short, in a terminal panicle. Heads *large, showy.*—Salt marshes and seashores.

ASTER.

A. Lindleya'nus, Torr. and Gr. Rather stout, smooth or sparsely pubescent. Leaves conspicuously serrate. Root-leaves and lowest stem-leaves ovate, more or less cordate, with margined petioles; uppermost *sessile, and pointed at both ends. Heads* *rather small, in a loose thyrse or panicle.* Scales linear, green-tipped. Rays pale violet.—Open barren grounds.

ACHILLEA.

A. Ptarmica, L. (SNEEZEWORT.) Leaves *simple*, lance-linear, with sharp appressed teeth. Rays 8-12, white, much longer than the bell-shaped involucre.

PRENANTHES.

P. Mainen'sis, Gray. Leaves resembling those of Nabalus racemosus, but the root-leaves ovate, and more abruptly narrowed to the short petiole. Heads 8-12-flowered, persistently drooping on slender pedicels.

ORDER LI. **CAMPANULA'CEÆ.** (CAMPANULA FAMILY.)

CAMPANULA.

C. rapunculoi'des, L. Flowers nodding, single in the axils of bracts, forming a raceme. Stem-leaves pointed, lanceolate, serrate; the lower cordate, long-petioled. (Int. from Eu.)

ORDER LII. **ERICACEÆ.** (HEATH FAMILY.)

GAYLUSSA'CIA.

G. dumo'sa, Torr. and Gr. (DWARF HUCKLEBERRY.) Fruit black, insipid. Racemes long with leaf-like, persistent bracts. Leaves obovate-oblong, mucronate.—Sandy low ground.

CALLU'NA.

C. vulga'ris, Salisb. (HEATHER.) A low evergreen shrub with numerous, opposite, minute leaves, mostly auricled at the base. Flowers axillary or terminating very short shoots, forming close racemes mostly one-sided, rose-colored, or white. Calyx of 4 sepals. Corolla 4-parted, bell-shaped. Calyx and corolla both persistent and becoming dry. Stamens 8. Capsule 4-celled.—Found sparingly in a few places on the coast of Nova Scotia and Newfoundland.

KAL'MIA.

K. angustifo'lia, L. (SHEEP-LAUREL, LAMBKILL.) Leaves opposite or in threes, oblong, obtuse, petioled. Corymbs *lateral, many-flowered.* Pod depressed. *Pedicels recurved in fruit.*—Bogs and damp barren grounds.

RHODODEN'DRON.

R. Rhodo'ra, Don. Corolla irregular, nearly an inch long, two-lipped; the upper lip 3-lobed; the lower lip of two oblong-linear, curved, nearly or quite distinct petals. Stamens 10, as long as the rose-colored corolla. Leaves alternate, oblong, somewhat pubescent. Shrub.—Bogs and damp barrens.

PYROLA.

P. mi'nor, L. Leaves roundish, slightly crenulate, thickish, usually longer than the margined petiole. *Raceme not one-sided.* Flowers white or rose-color. *Style short* and included in corolla.— Cold woods.

Order LIV.   PLUMBAGINACEÆ.   (Leadwort F.)
STAT'ICE.

S. Limo'nium, L. (Marsh Rosemary.) A maritime herb, with a thick, woody, astringent root, and oblong, spathulate or obovate-lanceolate radical leaves, tipped with a deciduous bristle. Flowers lavender-color, panicled on branching scapes. Calyx funnel-form, membranaceous. Corolla of 5 nearly or quite distinct petals, with the 5 stamens severally borne on their bases. Ovary 1-celled and 1-ovuled.—Salt marshes.

Order LV.   PRIMULA'CEÆ.   (Primrose Family.)
GLAUX.

G. marit'ima, L. (Sea Milkwort.) A fleshy herb, with usually opposite, oblong, entire, sessile leaves. Flowers single in the axils, nearly sessile. Calyx bell-shaped, 5-cleft, purplish and white. Corolla wanting. Stamens 5 on base of calyx. Capsule 5-valved, few-seeded.—Sea-shore.

Order LVI.   LENTIBULACEÆ.   (Bladderwort F.)
UTRICULA'RIA.

U. clandesti'na, Nutt. Stems and scapes slender. Leaves hair-like, bearing small bladders. Corolla yellow, lower lip 3-lobed, longer than the thick, blunt spur. Submersed stems bearing cleistogamous flowers.—Ponds.

ORDER LVIII. **SCROPHULARIA'CEÆ.** (FIGWORT F.)

VERON'ICA.

V. agres'tis, L. (FIELD SPEEDWELL.) Leaves round or ovate, crenate, petioled. Flowers in the axils of the ordinary leaves, long pedicelled. Seeds cup-shaped.—Sandy fields. (Int. from Eu.)

LINA'RIA.

L. Canaden'sis, Spreng. (WILD TOAD-FLAX.) A slender herb with linear, entire, alternate leaves. Flowers blue, small, in a naked, terminal raceme. Spur of corolla curved, filiform.—Sandy soil.

CASTILLE'IA.

C. pal'lida, Kunth., var. septentriona'lis, Gray. *Calyx equally cleft, divisions 2-cleft. Upper lip of corolla decidedly shorter than the tube.* Lower leaves linear; upper broader, mostly entire; the floral, oblong or obovate, greenish-white, varying to yellowish, purple or red.

PEDICULA'RIS.

P. Furbishiæ, Watson. Leaves pinnately-parted, and the short oblong divisions pinnately cut, or (in the upper) serrate. Calyx-lobes 5. Upper lip of corolla straight and beakless.—River banks.

ORDER LX. **LABIAT'Æ.** (MINT FAMILY.)

MEN'THA.

M. sati'va, L. (WHORLED MINT.) Flowers in globular clusters in the axils of leaves; the uppermost axils not flower-bearing. Leaves petioled, ovate, sharply serrate. Calyx with very slender teeth. (Int. from Eu.)

M. arvensis, L. (CORN MINT.) Flowers as in M. sativa, but leaves smaller, obtusely-serrate, and teeth of calyx short and broader. (Int. from Eu.)

ORDER LXI. **BORRAGINACEÆ.** (BORAGE FAMILY.)

MERTEN'SIA.

M. marit'ima, Don. (SEA-LUNGWORT.) Corolla white, trumpet-shaped, conspicuously 5-lobed, throat crested. Leaves fleshy, glaucous, ovate to spathulate. Stems spreading often decumbent.—Sea-beach.

Order LXIII. **POLEMONIA'CEÆ.** (Polemonium F.)

GIL'IA.

G. linea'ris, Gray. (Collo'mia linea'ris, Nutt.) A branching herb with alternate, linear-lanceolate, or oblong, sessile and entire leaves. Corolla salver-form, with stamens unequally inserted in its narrow tube, lilac-purple to nearly white. Ovules solitary.—Found on sands at the mouth of Eel River, Restigouche Co., N.B.

Order LXVI. **GENTIANA'CEÆ.** (Gentian Family.)

GENTIA'NA.

G. Amarel'la, L., var. acu'ta, Hook. f. Corolla somewhat funnel-form, mostly blue, its lobes entire, acute with a fringed crown at their base. Calyx lobes (4-5), lanceolate or linear, foliaceous.

G. linea'ris, Froel., var. latifolia, Gray. Flowers in a terminal cluster with a leafy involucre. Corolla blue, narrow funnel-form, with roundish-ovate lobes, and broad appendages. Leaves sessile, oblong-linear to ovate-lanceolate, smooth. Seeds winged.—Boggy places.

Order LXXII. **CHENOPODIACEÆ.** (Goosefoot F.)

SALICOR'NIA.

S. herbacea, L. (Samphire.) Flowers perfect in threes, embedded in hollows on the thickened upper joints, forming a spike. Calyx small and bladder-like, its margin toothed. Stamens 1 or 2.—Salt marshes. Often used as "greens."

SUÆ'DA.

S. linea'ris, Moq. (Suæda maritima, Gray.) (Sea-Blite.) A branching fleshy herb, with alternate, roundish, linear leaves. Flowers perfect, sessile in the axils of leafy bracts on slender branchlets. Sepals very thick. Stamens 5, with anthers exserted.

APPENDIX.

SALSO'LA.

S. Ka'li, L. (SALTWORT.) Flowers perfect, sessile, with two bractlets, single in axils of leaves. Calyx 5-parted, enclosing the depressed fruit. Stamens 5. A branching plant with alternate, awl-shaped, prickly-pointed leaves.—Sandy sea-shore.

ORDER LXXIV. **POLYGONA'CEÆ.** (BUCKWHEAT F.)

POLYG'ONUM.

P. maritimum, L. (COAST KNOTGRASS.) Prostrate with stout stems, glaucous. Leaves thick, oval to narrowly oblong. Flowers in the axils of leaves, clustered. Stipules very conspicuous. Stamens 8. Achenes smooth and shining, projecting above the calyx.—Sea-coast.

RU'MEX.

R. maritimus, L. (GOLDEN DOCK.) Low, slightly-pubescent, much branched. Leaves linear-lanceolate, wavy-margined, the lower auricled or heart-shaped at base. Flowers in whorls forming leafy spikes. Valves oblong, lance-pointed, each bearing 2-3 long bristles on each side, and a large grain on the back.—Sea-shore.

ORDER LXXVIII. **SANTALA'CEÆ.** (SANDALWOOD F.)

COMANDRA.

C. livida, Richardson. Peduncles axillary, slender, several-flowered. Leaves oval, alternate, almost sessile. Fruit pulpy when ripe, red.—Boggy barrens, near the coast.

ORDER LXXXII. **URTICA'CEÆ.** (NETTLE FAMILY.)

SUBORDER IV. **CANNABIN'EÆ.** (HEMP F.)

HU'MULUS.

H. Lupulus, L. (COMMON HOP.) A twining perennial. Leaves heart-shaped, mostly 3-5-lobed, petioled. Calyx of fertile flower a single sepal. In fruit the calyx, achene, etc., sprinkled with yellow resinous grains, which give the hop its taste and smell.

URTI'CA.

U. dioi'ca, L. (STINGING NETTLE.) Plant bristly with very stinging hairs. Leaves ovate, cordate, *very deeply serrate.* Spikes branching.—Waste places.

U. u'rens, L. Leaves elliptical or ovate, coarsely and deeply serrate with spreading teeth, petioled. Flower clusters 2 in each axil, composed of both staminate and pistillate flowers.—Waste grounds. (Int. from Eu.)

Order LXXXV. CUPULIF'ERÆ. (Oak Family.)
QUER'CUS.

Q. coccin'ea, Wang., var. ambig'ua, Gray. (GRAY OAK.) In this variety the leaves closely resemble those of Q. rubra, while the fruit is that of Q. coccinea.—Belleisle Bay, Kings Co., N.B.

Order LXXXVI. MYRICA'CEÆ. (Sweet-Gale F.)
MYRI'CA.

M. cerif'era, L. (BAYBERRY. WAX-MYRTLE.) Leaves oblong-lanceolate, entire, or wavy-toothed toward the apex, shining and sprinkled with resinous dots on both sides, fragrant. Sterile catkins scattered. Nuts naked, bony and covered with white wax.—Sandy soil near the coast.

Order LXXXVII. BETULA'CEÆ. (Birch Family.)
BET'ULA.

B. populifo'lia, Ait. (AMERICAN WHITE BIRCH. GRAY BIRCH.) Leaves very tremulous on slender petioles, triangular, very taper pointed, nearly truncate at base, smooth and shining except when young. Bark of trunk white, less separable than in Canoe Birch.—Poor soil.

Order LXXXVIII. SALICA'CEÆ. (Willow Family.)
SALIX.

S. fragilis, L. (CRACK WILLOW.) Leaves lanceolate, long pointed, pale or glaucous beneath, 3-6-inches long. Catkins borne on short, lateral, leafy branches. Stamens mostly 2, rarely 3-4 Capsule short-pedicelled. (Int. from Eu.)

S. balsamif′era, Barratt. A small much-branched shrub. Young twigs shining-chestnut on the sunny side. Leaves ovate-lanceolate, usually slightly cordate at base, at first very thin and of a reddish color, at length rigid, dark green above, and paler and conspicuously reticulate-veined beneath, slightly serrate, with slender petioles. Sterile catkins very silky with a few bracts at the base; fertile catkins leafy-peduncled, becoming very loose in fruit. Capsules long pedicelled.—Swamps.

S. myrtilloides, L. Low shrub. Leaves elliptic-obovate, entire, smooth, somewhat coriaceous when mature, revolute, reticulated, pale or glaucous beneath. Fertile catkins loosely-flowered on long leafy peduncles. Capsules glabrous, on slender pedicels.—Peat bogs.

ORDER LXXXVIII. (a.) **EMPETRA′CEÆ.** (CROW-BERRY FAMILY.)

EM′PETRUM.

E. ni′grum, L. (BLACK CROWBERRY.) A slender procumbent shrub with the foliage and aspect of a heath. Flowers polygamous, inconspicuous in axils of leaves. Calyx 3 petal-like sepals. Corolla wanting. Stamens 3. Fruit a blackberry-like drupe.—Sea-coast or near it.

CORE′MA.

C. Conrad′ii, Torr. (BROOM CROWBERRY.) Closely resembling the preceding. Flowers diœcious or polygamous, collected in terminal heads, each in the axil of a scaly bract, having no true calyx, but with 5 or 6 thin, dry bractlets under each. Stamens 3 or 4 with slender filaments. Drupes small, juiceless when ripe.

ORDER LXXXIX. **CONIF′ERÆ.** (PINE FAMILY.)

PI′NUS.

P. Banksia′na, Lambert. (GRAY OR NORTHERN SCRUB PINE.) Leaves in twos, about 1 inch long. Cones conical, usually curved, smooth and hard, about one and one-half inches long.—Barren soil.

JUNIPERUS.

J. Sabi'na, L., var. procumbens, Pursh. A procumbent or creeping shrub with two sorts of leaves, awl-shaped and scale-shaped, the latter acute. Fruit on short recurved peduncles.—Rocky banks and margins of swamps.

Order XCI.   **LEMNA'CEÆ.**   (Duckweed Family.)

LEM'NA.

L. trisul'ca, L. Fronds oblong-lanceolate, obscurely 3-nerved, sending off others from their sides which usually remain connected to them by slender stalks. Rootlets often absent.—Ponds and springy places.

Order XCII.   **TYPHA'CEÆ.**   (Cat-tail Family.)

SPARGA'NIUM.

S. min'imum, Fries. Usually floating, with very slender stems and flat narrow leaves. Stems shorter when growing out of water. Fertile flowers in 1 or 2 axillary heads. Fruit oblong-obovate, pointed, somewhat triangular.

Order XCIII.   **NAIADA'CEÆ.**   (Pondweed Family.)

POTAMOGETON.

P. Spiril'lus, Tuckerman. Stems very slender. Floating leaves when present oval to lanceolate, about as long as the petioles; submersed leaves narrowly-linear, or the upper ones broad-linear, or lance-oblong. Emersed flowers in many-flowered spikes; submersed flowers usually solitary. Fruit either winged and 4-5-toothed, or wingless and entire.

P. prælon'gus, Wulfen. Stem very long and branching. Leaves all submersed and similar, lanceolate, half-clasping, with a boat-shaped cavity at the end. Spikes loose-flowered with very long peduncles. Fruit sharply keeled when dry. Stem white.—Ponds and large rivers.

P. Robbin'sii, Oakes. Leaves all submersed and similar, narrowly lanceolate or linear, crowded in 2 ranks, recurved, serrulate, many nerved. Stems rigid with numerous branches. Fruit keeled with a broadish wing.—Lakes and slow streams.

ORDER XCIV.   **ALISMA'CEÆ.**   (WATER-PLANTAIN F.)

SAGITTA'RIA.

S. calyci'na, Engelm., var. spongiosa, Engelm. Scape weak, and at length usually procumbent. Fertile flowers perfect. Leaves broadly halberd-shaped with wide-spreading lobes. Submerged leaves without blades.

ORDER XCVI.   **ORCHIDA'CEÆ.**   (ORCHIS FAMILY.)

SPIRAN'THES.

S. latifo'lia, Torr. Flowers white, in 3 ranks, forming a narrow spike. Lip oblong, yellowish on the face, not contracted in the middle, wavy-crisped at the blunt apex. Stem nearly naked. Leaves oblong or lance-oblong.—Moist banks.

HABENA'RIA.

H. fimbria'ta, R. Br. Resembling H. psycodes, but the flowers 3 or 4 times larger, and the petals toothed down the sides. Divisions of the large lip more fringed. Spike loosely flowered.—Wet meadows.

ORDER CI.   **LILIACEÆ.**   (LILY FAMILY.)

STREP'TOPUS.

S. amplexifo'lius, DC. Differs from S. roseus in that the leaves are smooth and glaucous underneath, (instead of being green and finely ciliate); and the branches glabrous, (instead of being beset with bristly hairs). Flower greenish-white on a long abruptly-bent peduncle.—Cold moist woods.

AL'LIUM.

A. Schœnop'rasum, L. (CHIVES.) Leaves linear, hollow. Scape naked, or leafy at the base. Flowers rose-purple, in a globular umbel. Sepals lanceolate, pointed. *Ovary not crested.*—Margins of rivers.

APPENDIX.

Order CII. **JUNCA'CEÆ.** (Rush Family.)

JUN'CUS.

J. Styg'ius, L. Scape slender, 1-3-leaved below, naked above. Leaves thread-like, hollow, not knotted. Heads 1 or 2, of 3-4 flowers, about as long as the dry, awl-pointed sheathing bract. Stamens 6.—Peat bogs.

J. Greenii, Oakes and Tuckerm. Stems slender, simple, tufted. Leaves nearly terete, deeply channelled on the inner side. Flowers solitary, panicled. The principal leaf of the involucre usually much longer than the panicle. Pod ovoid-oblong, obtuse, longer than the acute sepals. Seeds ribbed, and delicately cross-lined.—Sandy ground on or near the coast.

Order CVII. **FIL'ICES.** (Fern Family.)

ASPLE'NIUM.

A. vi'ride, Hudson. Resembling A. Trichomanes, but less rigid, and the stipe brownish at base, becoming green upwards. Pinnæ roundish-ovate or ovate-rhomboid, short-stalked, crenately toothed.—Shaded cliffs.

PHEGOP'TERIS.

P. calca'rea, Fée. Closely resembling P. Dryopteris, but differs in the fronds being *minutely glandular and somewhat rigid*, and in the lowest pinnæ on the lower side of the lateral divisions *proportionally smaller.*—Rocky hillsides, Restigouche River.

ASPID'IUM.

A. fra'grans, Swartz. Fronds 4-12 inches high, fragrant, narrowly lanceolate, with narrowly-oblong pinnately-parted pinnæ, their divisions nearly covered beneath by very large thin indusia.—Rocks, Restigouche River, near mouth of Patapedia, and at the railway tunnel, Restigouche.

A. Fi'lix-mas, Swartz. (Male-Fern.) Fronds lanceolate, very chaffy at the base, twice pinnate except that the upper pinnules run together. Pinnæ linear-lanceolate, tapering from the base to the summit. Pinnules very obtuse, the basal ones incisely lobed. Fruit dots rather closer to the midvein than the margin. Indusium convex, persistent.—Rocky woods.

A. aculea'tum, Swartz, var., Braun'ii Koch. Fronds twice pinnate, oblong-lanceolate, narrowing gradually toward the base. Pinnules ovate or oblong, truncate and nearly rectangular at the base (the lower short-stalked), beset with long and soft as well as chaffy hairs.—Indusium fixed by the centre. Fronds evergreen, very chaffy on the stalk and rhachis.—Ravines and deep woods.

WOOD'SIA.

Fruit dots on the back of free veins, circular, with a very thin indusium fastened by its base all around under the spore-cases. Small tufted ferns with pinnately-divided fronds.

W. Ilven'sis, R. Brown. Stalks indistinctly jointed at some distance above the base. Fronds oblong-lanceolate, 2-6 inches long, rather smooth and green above, *thickly clothed below with bristly rusty chaff*, pinnate; the pinnæ oblong, blunt, sessile, pinnately-parted with the segments indistinctly crenate. Fruit dots near the margin.—Exposed rocks.

W. hyperbo'rea, R. Brown. Stalks jointed. Fronds narrowly oblong-lanceolate. *Sparingly hairy beneath with chaffy hairs*. pinnate; the pinnæ triangular-ovate, pinnately lobed, *the lobes few and almost entire*.—Ravines.

W. glabel'la. R. Brown. Stalks jointed as in the two preceding species. Frond *linear*, very delicate, *smooth*, pinnate. Pinnæ roundish-ovate, the lower somewhat distant, *crenately-lobed*.—Moist rocks.

W. obtu'sa, Torr. *Stalks not jointed*. Frond broadly lanceolate, beset with small glandular hairs, once or nearly twice pinnate. Pinnæ pinnately parted. Segments of pinnæ crenately toothed.—Cliffs and rocky places.

BOTRYCH'IUM.

B. lanceola'tum, Angstrœm. Frond 3-10 inches high. *Sterile part closely sessile at the top* of the slender common stalk, very slightly fleshy, triangular, ternately twice pinnatifid, with *acute*, toothed lobes. Veinlets branching from the continuous midvein. The fertile part twice or thrice pinnate.

B. matricariæfo'lium, Braun. Resembling the preceding, but the sterile segment not quite sessile, somewhat fleshy, pinnate to twice pinnatifid, with *obtuse* lobes. Midvein *broken up* into forking veinlets.—Damp woods.

B. sim'plex, Hitchcock. Fronds small, seldom 6 inches high, *the sterile segment borne near the middle of the plant*, short petioled, fleshy, simple, and roundish, or pinnately lobed, with roundish lobes decurrent on the broad and flat indeterminate rhachis. *Veins all forking from the base.*

Order CVIII. **EQUISETA'CEÆ.** (Horsetail Family.)

EQUISE'TUM.

E. praten'se, Ehrh. Stems more slender, with *3-sided simple branches* shorter than in E. arvense. Stem-sheaths with short, ovate-lanceolate teeth (those of the branches 3-toothed.) The fertile stems produce branches when older, except at the top, which perishes after fructification.—Low meadows.

E. sylvat'icum, L. *Branches compound with loose sheaths*, those of the stem having 8-14 somewhat blunt teeth, while those of the branches have 4-5 (of the branchlets 3) lance-pointed diverging teeth. Top withering away after fructification.—Damp shady places.

Order CIX. **LYCOPODIA'CEÆ.** (Club-Moss Family.)

LYCOPO'DIUM.

L. Sela'go, L. Resembling L. lucidulum, but the stems more rigid, and the leaves ascending and all alike, while in L. lucidulum they consist of alternate zones of shorter and longer leaves.

L. inunda'tum, L. A low plant with weak, creeping, sterile stems, and solitary erect fertile stems bearing a short, thick, leafy spike. Stem-leaves lanceolate, acute, soft, spreading; those of the spike closely resembling them.—Sandy bogs.

ISO'ETES.

I. echinos'pora, Durieu, var. Braun'ii, Engelm. (Quillwort.) A small aquatic grass-like plant with a corm-like stem, bearing 15-30 slender leaves. The large sporangia axillary, partly enwrapped by the thin edges of the excavated bases of the leaves, beset with small spinules.—Lakes and ponds.

INDEX.

The names of the Orders, Classes, and Divisions are in large capitals; those of the Sub-orders in small capitals. The names of Genera, as well as popular names and synonyms, are in ordinary type.

	PAGE.		PAGE.
Abies	141	ANACARDIACEÆ	28
ABIETINEÆ	140	Anagallis	93
Abutilon	25	Andromeda	89
Acalypha	126	Anemone	3
Acer	31	ANGIOSPERMS	1
Achillea	80	ANONACEÆ	7
Acorus	144	Antennaria	71
Actæa	6	APETALOUS EXOGENS	116
Adam-and-Eve	155	Aphyllon	94
Adder's-Mouth	154	Apios	38
Adder's-Tongue	181	Aplectrum	155
Adiantum	176	APOCYNACEÆ	114
Adlumia	11	Apocynum	114
Agrimonia	41	Apple	45
Agrimony	41	AQUIFOLIACEÆ	90
Alder	136	Aquilegia	6
Alisma	148	Arabis	14
ALISMACEÆ	147	ARACEÆ	143
Alkanet	107	Aralia	56
Allium	162	ARALIACEÆ	56
Alnus	136	Arbor Vitæ	141
Alyssum	15	Archangelica	55
Amaranth	118	Archemora	55
Amaranth Family	118	Arctostaphylos	88
Amarantus	118	Arenaria	22
AMARANTACEÆ	118	Arethusa	154
AMARYLLIDACEÆ	156	Arisæma	144
Amaryllis Family	156	ARISTOLOCHIACEÆ	116
Ambrosia	70	Aromatic Wintergreen	88
Amelanchier	45	Arrow-Grass	147
American Columbo	112	Arrow-Head	148
American Laurel	89	Arrow-Wood	60
Ampelopsis	29	Artemisia	70
Amphicarpæa	38	ARTOCARPEÆ	127
AMYGDALEÆ	38	Arum Family	143
Anacharis	148	Asarum	116

INDEX

	PAGE		PAGE
ASCLEPIADACEÆ	114	Bittersweet	111
Asclepias	114	Black Alder	90
Ash	11	Blackberry	44
Asimina	7	Black Bindweed	121
Aspen	139	Black-Mustard	15
Aspidium	178	Black Snake-root	54
Asplenium	177	Bladder Campion	21
Aster	75	Bladder Fern	179
Astragalus	35	Bladder-Nut	31
Atriplex	118	Bladderwort	93
Avens	41	Bladderwort Family	93
Balm of Gilead	139	Blazing-Star	71
Balsam Family	27	Blite	118
BALSAMINACEÆ	27	Blitum	118
Baneberry	6	Blood-root	10
Baptisia	38	Blue Beech	134
Barberry Family	8	Blueberry	87
Barren Strawberry	42	Bluebottle	69
Basil	10, 104	Blue Cohosh	8
Basswood	25	Blue Flag	156
Bastard Toad-flax	124	Blue-eyed Grass	156
Bayberry	135	Blue Lettuce	83
Beach Pea	37	Blue-weed	106
Bearberry	88	Bœhmeria	129
Beard-Tongue	97	Boneset	72
Beaver-Poison	55	Borage Family	105
Beech	134	BORRAGINACEÆ	105
Beech-drops	94	Botrychium	180
Beech-Fern	178	Bouncing Bet	21
Bedstraw	62	Bowman's Root	41
Beggar's Lice	107	Bracted Bindweed	110
Beggar-ticks	79	Bracken	176
Bellflower	84	Brake	176
Bellwort	160	Bramble	43
BERBERIDACEÆ	8	Brasenia	9
Bergamo	104	Brassica	15
Betula	136	Bristly Sarsaparilla	57
BETULACEÆ	135	Brooklime	96
Bidens	79	Brook-weed	93
Bindweed	110, 121	Broom-rape Family	94
Birch	136	Brunella	104
Birch Family	135	Buckbean	113
Birthwort Family	116	Buckthorn	29
Bishop's-Cap	47	Buckthorn Family	29
Bitter-Cress	14	Buckwheat	122
Bitter-Nut	131	Buckwheat Family	119

	PAGE
Bugbane	6
Bugseed	118
Bugle-weed	103
Bunch-berry	57
Burdock	69
Bur-Marigold	79
Burning-Bush	30
Bur-reed	145
Bush-Clover	36
Bush-Honeysuckle	60
Butter-and-Eggs	97
Buttercup	4
Butterfly-weed	115
Butternut	130
Butter-weed	77
Butterwort	94
Button-bush	62
Buttonwood	130
Cakile	16
Calamintha	104
Calaminth	104
Calamus	144
Calla	144
Calopogon	154
Caltha	5
Calypso	154
Calystegia	110
Camelina	15
Campanula	84
CAMPANULACEÆ	84
Campanula Family	84
Campion	21
Camptosorus	178
CANNABINEÆ	128
Cannabis	129
Caper Family	16
CAPPARIDACEÆ	16
CAPRIFOLIACEÆ	58
Capsella	16
Carex	168
Cardamine	14
Cardinal Flower	83
Carpet-weed	52
Carpinus	134

	PAGE
Carrion Flower	158
Carrot	54
Carya	131
CARYOPHYLLACEÆ	21
Cashew Family	28
Cassandra	88
Castanea	134
Castilleia	98
Catbrier	157
Catchfly	21
Catmint	104
Catnip	104
Cat-tail Family	144
Cat-tail Flag	145
Caulophyllum	8
Ceanothus	30
Cedar	141
Celandine	10
CELASTRACEÆ	30
Celastrus	30
Celtis	129
Centaurea	69
Cephalanthus	62
Cerastium	23
CERATOPHYLLACEÆ	124
Ceratophyllum	124
Chain-Fern	177
Charlock	15
Chelidonium	10
Chelone	97
CHENOPODIACEÆ	116
Chenopodium	117
Cherry	39
Chestnut	134
Chickweed	23
Chickweed-Wintergreen	92
Chimaphila	89
Chiogenes	88
Choke-berry	43
Choke-Cherry	40
Chrysoplenium	48
Cichorium	81
Cichory	81
Cicuta	55
Cimicifuga	6
Cinnamon Fern	180

INDEX

	PAGE.		PAGE.
Cinque-foil	42	Cowbane	55
Circæa	50	Cow-Parsnip	54
Cirsium	68	Cowslip	92
CISTACEÆ	18	Cow-Wheat	99
Claytonia	24	Crab-Apple	45
Clearweed	129	Cranberry	87
Cleavers	62	Cranberry-tree	61
Clematis	3	Cranesbill	26
Cliff-brake	176	CRASSULACEÆ	48
Climbing Bittersweet	30	Cratægus	45
Clintonia	160	Creeping-Snowberry	88
Clotbur	69	Cress Family	12
Clover	34	Crowfoot	4
Club-Moss	183	Crowfoot Family	2
Club-Moss Family	182	CRUCIFERÆ	12
Cockle	22	CRYPTOGAMS	169
Cocklebur	69	Cryptotænia	56
Cohosh	8	Cuckoo-flower	14
Collinsonia	103	CUCURBITACEÆ	52
Columbine	6	Cudweed	71
Comandra	124	Cup-plant	81
Comfrey	106	CUPRESSINEÆ	140
COMPOSITÆ	64	CUPULIFERÆ	131
Composite Family	64	Currant	46
Comptonia	135	Cuscuta	110
Cone-Flower	79	Custard-Apple Family	7
CONIFERÆ	139	Cynoglossum	107
Conioselinum	55	Cynthia	81
Conium	56	CYPERACEÆ	165
Conopholis	94	Cyperus	166
CONVOLVULACEÆ	109	Cypripedium	155
Convolvulus	110	Cystopteris	179
Convolvulus Family	109		
Coptis	6	Dalibarda	43
Corallorhiza	154	Dandelion	82
Coral-root	154	Datura	112
Corispermum	118	Daucus	54
CORNACEÆ	57	Deer-Grass	51
Corn-Cockle	22	Dentaria	13
Cornel	57	Desmodium	35
Cornus	57	Dewberry	44
Corpse-Plant	90	Dicentra	11
Corydalis	11	Dicksonia	180
Corylus	134	DICOTYLEDONS	1
Cottonwood	139	Diervilla	60

	PAGE		PAGE
Dioscorea	157	Eriocaulon	165
DIOSCOREACEÆ	157	ERIOCAULONACEÆ	165
Diplopappus	78	Eriophorum	167
DIPSACEÆ	63	Erodium	26
Dipsacus	63	Erythronium	162
Dirca	123	Erysimum	15
Ditch Stone-crop	48	Euonymus	30
Dock	121	Eupatorium	72
Dockmackie	61	Euphorbia	125
Dodder	110	EUPHORBIACEÆ	125
Dogbane	114	Euphrasia	99
Dogbane Family	114	Evening Primrose	50
Dog's-tooth Violet	162	Evening Primrose Family	49
Dogwood	57	Everlasting	71
Dogwood Family	57	Everlasting Pea	37
Double-bristled Aster	78	EXOGENS	1
Draba	15	Eyebright	99
Drosera	19		
DROSERACEÆ	19	Fagopyrum	122
Duckweed	144	Fagus	134
Duckweed Family	144	Fall Dandelion	81
Dutchman's Breeches	11	False Asphodel	160
		False Dragon-head	104
Echinocystis	52	False Flax	15
Echinospermum	106	False Gromwell	107
Echium	106	False Indigo	38
Eel-Grass	149	False Lettuce	83
ELÆAGNACEÆ	123	False Loosestrife	51
Elder	60	False Mitre-Wort	47
Elecampane	72	False Nettle	129
Eleocharis	167	False Pennyroyal	102
Elm	128	False Pimpernel	98
Elm Family	127	False Solomon's Seal	161
Elodes	21	False Spikenard	161
ENDOGENS	143	FERNS	169-174
Enchanter's Nightshade	50	Fever-bush	123
Epigæa	88	Fever-wort	60
Epilobium	50	FICOIDEÆ	52
Epiphegus	94	Figwort	97
EQUISETACEÆ	181	Figwort Family	94
Equisetum	181	Filbert	134
Erechthites	71	FILICES	174
ERICACEÆ	85	Fir	141
ERICINEÆ	86	Fire-Pink	22
Erigenia	56	Fireweed	71
Erigeron	77	Five-Finger (Cinque-Foil)	42

INDEX.

	PAGE
Flax	25
Flax Family	25
Fleabane	77
Floating-Heart	114
Flower-de-Luce	156
Flowering Fern	180
FLOWERING PLANTS	1
FLOWERLESS PLANTS	169
Forget-me-not	108
Fragaria	43
Frasera	112
Fraxinus	115
Frog's-bit Family	148
Frostweed	18
FUMARIACEÆ	11
Fumitory	11
Fumitory Family	11
Galeopsis	105
Galium	62
GAMOPETALOUS EXOGENS	53
Garlic	162
Gaultheria	88
Gaylussacia	87
Gentian	113
Gentiana	113
GENTIANACEÆ	112
Gentian Family	112
GERANIACEÆ	26
Geranium	26
Geranium Family	26
Gerardia	98
Germander	102
Geum	41
Giant-Hyssop	104
Gillenia	41
Ginseng	56
Ginseng Family	56
Gleditschia	38
Gnaphalium	71
Goat's-Beard	83
Golden-Rod	72
Golden Saxifrage	48
Gold-Thread	6
Goodyera	153
Gooseberry	46

	PAGE
Goosefoot	117
Goosefoot Family	116
Goose-Grass	62, 119
Gourd Family	52
GRAMINEÆ	168
Grape	29
Grass Family	168
Grass of Parnassus	47
Gratiola	98
Great Angelica	55
Greenbrier	157
Green Dragon	144
Gromwell	107
Ground Cherry	111
Ground Hemlock	142
Ground Ivy	104
Ground Laurel	88
Ground-nut	38
Ground-Pine	183
Groundsel	72
GYMNOSPERMS	139
Habenaria	151
Hackberry	129
Halenia	113
HALORAGEÆ	49
HAMAMELACEÆ	48
Hamamelis	48
Harbinger-of-Spring	56
Harebell	84
Hart's-Tongue	177
Hawkweed	81
Hawthorn	45
Hazel-nut	134
Heal-all	104
Heath Family	85
Hedeoma	103
Hedge Bindweed	110
Hedge-Hyssop	98
Hedge-Mustard	15
Hedge-Nettle	105
Helenium	78
Helianthemum	18
Helianthus	79
Heliopsis	80
Hemlock	141

INDEX.

	PAGE		PAGE
Hemlock-Parsley	55	Hypopitys	90
Hemlock-Spruce	141	Hypoxys	157
Hemp	120		
Hemp Family	128	Ice-Plant Family	52
Hemp-Nettle	105	Ilex	90
Henbane	111	Ilysanthes	98
Hepatica	3	Impatiens	27
Heracleum	54	Indian Cucumber-root	160
Herb-Robert	26	Indian Hemp	114
Hickory	131	Indian Mallow	25
Hieracium	81	Indian Physic	41
Hippuris	49	Indian Pipe	90
Hoary Puccoon	107	Indian Tobacco	84
Hobble-bush	61	Indian Turnip	144
Hog Pea-nut	38	Inula	72
Hogweed	70	IRIDACEÆ	155
Holly	90	Iris	156
Holly Family	90	Iris Family	155
Honey-Locust	38	Iron-wood	134
Honeysuckle	59	Isanthus	102
Honeysuckle Family	58		
Honewort	56	Jeffersonia	8
Hop-Hornbeam	134	Jerusalem Artichoke	70
Horehound	105	Jerusalem Oak	117
Hornbeam	134	Jewel-Weed	27
Hornwort	124	Joe-Pye Weed	72
Hornwort Family	124	Juglans	130
Horse-Balm	103	JUGLANDACEÆ	130
Horse-Mint	103	JUNCACEÆ	162
Horseradish	13	Juncus	163
Horsetail Family	181	June-berry	45
Horse-weed	77	Juniper	141
Hound's-Tongue	107	Juniperus	141
Houstonia	63		
Huckleberry	87	Kalmia	89
Hudsonia	19	Knotgrass	119
Huntsman's Cup	10	Knotweed	119
Hydrastis	6		100
HYDROCHARIDACEÆ	148	Labrador Tea	86
Hydrocotyle	54	Lactuca	82
HYDROPHYLLACEÆ	108	Lady's Slipper	155
Hydrophyllum	108	Lady's Thumb	120
Hyoscyamus	111	Lady's Smock	14
HYPERICACEÆ	19	Ladies' Tresses	153
Hypericum	20	Lake-Cress	13

INDEX.

	PAGE.		PAGE.
Lamb's Quarters	117	LOBELIACEÆ	83
Lampsana	81	Lobelia Family	83
Laportea	129	Locust-tree	35
Lappa	69	Lonicera	59
Larch	141	Looser' &c.	92
Larix	141	Loosestrife Family	51
Lathyrus	37	Lophanthus	104
LAURACEÆ	122	Lopseed	100
Laurel Family	122	Lousewort	9(
Laurestinus	60	Lucerne	35
Leaf-Cup	80	Ludwigia	51
Leather-leaf	88	Lupine	34
Leatherwood	123	Lupinus	34
Lechea	19	Luzula	163
Ledum	89	Lychnis	22
Leek	162	Lycium	111
LEGUMINOSÆ	33	LYCOPODIACEÆ	182
LEMNACEÆ	144	Lycopodium	183
LENTIBULACEÆ	93	Lycopus	103
Leontodon	81	Lysimachia	92
Leonurus	105	LYTHRACEÆ	51
Lepidium	16		
Lespedeza	36	Madder Family	61
Lettuce	82	MAGNOLIACEÆ	6
Leucanthemum	78	Magnolia Family	6
Liatris	71	Maidenhair	176
LIGULIFLORÆ	67	Mallow	24
LILIACEÆ	158	Mallow Family	24
Lilium	161	Malva	24
Lily	161		24
Lily Family	158	Mandrake	8
Limnanthemum	114	Maple	31
LINACEÆ	25	Mare's-Tail	49
Linaria	97	Marrubium	105
Linden Family	25	Marsh-Cress	13
Lindera	123	Marsh-Marigold	5
Linnæa	59	Marsh St. John's-wort	21
Linum	25	Maruta	78
Liparis	154	Matrimony-Vine	111
Liriodendron	7	May-Apple	8
Listera	153	Mayflower	88
Lithospermum	107	Mayweed	78
Liver-leaf	3	Meadow-Beauty	51
Lizard's-tail	124	Meadow-Parsnip	55
Lizard's-tail Family	124	Meadow-Rue	4
Lobelia	83	Meadow-Sweet	40

INDEX.

	Page		Page
Medeola	160	Mountain Mint	103
Medicago	35	Mouse-ear Chickweed	23
Medick	35	Mugwort	70
Melampyrum	99	Mulberry	129
Melastoma Family	51	Mulgedium	83
MELASTOMACEÆ	51	Mullein	96
Melilot	35	Musk-Mallow	25
Melilotus	35	Mustard	15
MENISPERMACEÆ	7	Myosotis	108
Menispermum	7	Myrica	135
Mentha	102	MYRICACEÆ	134
Menyanthes	113	Myriophyllum	49
Mexican Tea	117		
Mezereum Family	123	Nabalus	82
Microstylis	154	NAIADACEÆ	145
Milfoil	80	Naked Broom-rape	94
Milk-Vetch	35	Nardosmia	75
Milkweed	114	Nasturtium	13
Milkweed Family	114	Neckweed	97
Milkwort	32	Nemopanthes	90
Milkwort Family	32	Nepeta	104
Mimulus	97	Nesæa	51
Mint	102	Nettle	129
Mint Family	100	Nettle Family	127
Mitchella	63	Nettle-tree	129
Mitella	47	New Jersey Tea	30
Mitrewort	47	Nicotiana	112
Mocassin Flower	155	Nightshade	111
Mollugo	52	Nightshade Family	110
Monarda	103	Nine-Bark	40
Moneses	89	Nipple-wort	81
Monkey-Flower	97	Nuphar	9
MONOCOTYLEDONS	143	Nymphæa	9
Monotropa	90	NYMPHÆACEÆ	9
MONOTROPEÆ	87		
Montelia	118	Oak	132
Moonseed	7	Oak Family	131
Moonseed Family	7	Œnothera	50
Moonwort	180	OLEACEÆ	115
Moosewood	123	Oleaster Family	123
Morus	129	Olive Family	115
Mossy Stone-crop	48	ONAGRACEÆ	49
Motherwort	105	Onion	162
Mountain Ash	45	Onoclea	180
Mountain Holly	90	Onopordon	60
Mountain Maple	31	Onosmodium	107

INDEX.

	PAGE.		PAGE.
OPHIOGLOSSACEÆ	176	Physostegia	104
Ophioglossum	181	Phytolacca	116
Orache	118	PHYTOLACCACEÆ	116
Orange-root	6	Pickerel-weed	164
ORCHIDACEÆ	149	Pickerel-weed Family	164
Orchis	151	Pignut	131
Orchis Family	149	Pigweed	117, 118
OROBANCHACEÆ	94	Pilea	129
Orpine	48	Pimpernel	93
Orpine Family	48	Pine	140
Osmorrhiza	56	Pine-drops	96
OSMUNDACEÆ	176	Pine Family	139
Osmunda	180	Pine-sap	96
Ostrich Fern	180	Pinguicula	94
Ostrya	134	Pink Family	21
Oswego Tea	103	Pinus	140
OXALIDACEÆ	27	Pinweed	19
Oxalis	27	Pipewort	165
Ox-Eye	80	Pipewort Family	165
Ox-eye Daisy	78	Pipsissewa	86
		Pitcher-Plant Family	10
Painted-Cup	98	Plane-tree	136
PAPAVERACEÆ	10	Plane-tree Family	130
Papaw	7	PLANTAGINACEÆ	91
Parietaria	129	Plantago	91
Parnassia	47	Plantain	91
Parsley Family	53	Plantain Family	91
Parsnip	55	PLATANACEÆ	130
Partridge-berry	63	Platanus	130
Pastinaca	55	Pleurisy-root	115
Pear	45	Plum	39
Pedicularis	99	Podophyllum	8
Pellæa	176	Pogonia	154
Pellitory	129	Poison Elder	28
Pennycress	16	Poison Hemlock	56
Pennyroyal	103	Poison Ivy	28
Penthorum	48	Pokeweed	116
Pentstemon	97	Pokeweed Family	116
Peppergrass	16	Polanisia	17
Peppermint	102	POLEMONIACEÆ	109
Pepper-root	13	Polemonium Family	109
PHANEROGAMS	1	Polygala	32
Phegopteris	178	POLYGALACEÆ	32
Phlox	109	POLYGONACEÆ	119
Phryma	100	Polygonatum	161
Physalis	111	Polygonum	119

194 INDEX.

	PAGE.		PAGE.
Polymnia	80	Rattlesnake-Plantain	153
POLYPETALOUS EXOGENS	1	Rattlesnake-root	82
POLYPODIACEÆ	174	Rattlesnake-weed	82
Polypodium	176	Rein-Orchis	151
Polypody	176	RHAMNACEÆ	29
POMEÆ	39	Rhamnus	29
Pondweed	146	Rhexia	51
Pondweed Family	145	Rhinanthus	99
Pontederia	164	Rhus	28
PONTEDERIACEÆ	164	Ribes	46
Poplar	139	Rib-grass	91
Poppy Family	10	Rich-weed	103, 129
Populus	139	Robinia	35
Portulaca	24	Robin's-Plantain	77
PORTULACACEÆ	23	Rock-Cress	14
Potamogeton	146	Rock-Rose	18
Potentilla	42	Rock-Rose Family	18
Prairie Dock	81	Rosa	44
Prickly Ash	28	Rose	44
Primrose	92		38
Primrose Family	91	Rose Family	38
Primula	92	Rosin Plant	81
PRIMULACEÆ	91	RUBIACEÆ	61
Prince's Pine	89	Rubus	43
Prosartes	160	Rudbeckia	79
Prunus	39	Rue Family	27
Pteris	176	Rumex	121
Pterospora	90	Rush	162
Puccoon	107	Rush Family	162
Pulse Family	33	RUTACEÆ	27
Purslane	24		
Purslane Family	23	Sagittaria	148
Putty Root	155	St. John's-wort	20
Pycnanthemum	103	St. John's-wort Family	19
Pyrola	89	SALICACEÆ	136
PYROLEÆ	86	Salix	137
Pyrus	45	Salsify	83
		Salsola	117
Quercitron	133	Sambucus	60
Quercus	132	Samolus	93
		Sandalwood Family	124
Ragweed	70	Sandwort	22
Ragwort	72	Sanguinaria	10
RANUNCULACEÆ	2	Sanicle	54
Ranunculus	4	Sanicula	54
Raspberry	43	SANTALACEÆ	124

www.ingramcontent.com/pod-product-compliance
Lightning Source LLC
Chambersburg PA
CBHW031742230426
43669CB00007B/451